Controlling Energy Demand in Mobile Computing Systems

Synthesis Lectures on Mobile and Pervasive Computing

Editor
Mahadev Satyanarayanan, *Carnegie Mellon University*

Mobile computing and pervasive computing represent major evolutionary steps in distributed systems, a line of research and development that dates back to the mid-1970s. Although many basic principles of distributed system design continue to apply, four key constraints of mobility have forced the development of specialized techniques. These include: unpredictable variation in network quality, lowered trust and robustness of mobile elements, limitations on local resources imposed by weight and size constraints, and concern for battery power consumption. Beyond mobile computing lies pervasive (or ubiquitous) computing, whose essence is the creation of environments saturated with computing and communication, yet gracefully integrated with human users. A rich collection of topics lies at the intersections of mobile and pervasive computing with many other areas of computer science.

RFID Explained
Roy Want
2006

Controlling Energy Demand in Mobile Computing Systems
Carla Schlatter Ellis
2007

Controlling Energy Demand in Mobile Computing Systems
Carla Schlatter Ellis

ISBN: 978-3-031-01347-8   paperback
ISBN: 978-3-031-02475-7   ebook

DOI 10.1007/978-3-031-02475-7

A Publication in the Springer series

SYNTHESIS LECTURES IN MOBILE AND PERVASIVE COMPUTING #2

Lecture #2

Series Editor: Mahadev Satyanarayanan, Carnegie Mellon University

Library of Congress Cataloging-in-Publication Data

Series ISSN: 1933-9011   print
Series ISSN: 1933-902X   electronic

First Edition
10 9 8 7 6 5 4 3 2 1

Controlling Energy Demand in Mobile Computing Systems

Carla Schlatter Ellis
Duke University

SYNTHESIS LECTURES IN MOBILE AND PERVASIVE COMPUTING #2

ABSTRACT

This lecture provides an introduction to the problem of managing the energy demand of mobile devices. Reducing energy consumption, primarily with the goal of extending the lifetime of battery-powered devices, has emerged as a fundamental challenge in mobile computing and wireless communication. The focus of this lecture is on a systems approach where software techniques exploit state-of-the-art architectural features rather than relying only upon advances in lower-power circuitry or the slow improvements in battery technology to solve the problem. Fortunately, there are many opportunities to innovate on managing energy demand at the higher levels of a mobile system. Increasingly, device components offer low power modes that enable software to directly affect the energy consumption of the system. The challenge is to design resource management policies to effectively use these c apabilities.

The lecture begins by providing the necessary foundations, including basic energy terminology and widely accepted metrics, system models of how power is consumed by a device, and measurement methods and tools available for experimental evaluation. For components that offer low power modes, management policies are considered that address the questions of when to power down to a lower power state and when to power back up to a higher power state. These policies rely on detecting periods when the device is idle as well as techniques for modifying the access patterns of a workload to increase opportunities for power state transitions. For processors with frequency and voltage scaling capabilities, dynamic scheduling policies are developed that determine points during execution when those settings can be changed without harming quality of service constraints. The interactions and tradeoffs among the power management policies of multiple devices are discussed. We explore how the effective power management on one component of a system may have either a positive or negative impact on overall energy consumption or on the design of policies for another component. The important role that application-level involvement may play in energy management is described, with several examples of cross-layer cooperation. Application program interfaces (APIs) that provide information flow across the application-OS boundary are valuable tools in encouraging development of energy-aware applications. Finally, we summarize the key lessons of this lecture and discuss future directions in managing energy demand.

KEYWORDS

Energy, power management, battery power, voltage scaling, wireless, operating system

Contents

CHAPTER 1

Introduction

Managing energy consumption, primarily with the goal of extending the lifetime of battery-powered devices, is widely recognized as a fundamental challenge in mobile computing and wireless communication. The limited availability of battery energy has emerged as a significant problem with the growth in reliance on mobile technology, the increased capabilities of the hardware, and the rising expectations of users for ubiquitous services.

Mobility implies that a device carries its own energy supply, typically in the form of batteries. Weight and form factor play a role in limiting the energy capacity available. Lithium-ion batteries have become the dominant choice for rechargeable batteries in mobile electronics because of their light weight, high energy density, and memory-free recharge characteristics. However, lithium-ion battery density is only improving at a rate of approximately 10% per year (Fig. 1.1) while demand for more features and greater performance can easily eat up that improvement, resulting in no real increase in runtime that can be attributed to advances in battery technology. The fact that batteries have become a major headache for users can be observed at any airport where travelers focus on finding electrical outlets to recharge their laptops and that low batteries have become a widely accepted excuse for truncated cell phone conversations.

The focus of this lecture is on managing the energy *demand* of mobile devices through a *systems* approach. The supply side of the problem offers important and interesting topics including battery modeling, interfacing with smart batteries, exploiting the chemical characteristics of batteries, harvesting energy from the environment (e.g., solar, kinetic), and developing alternative mobile energy sources. Exploration of the energy supply issues for mobile computing will be the topic of a separate lecture.

The second aspect of our focus is the systems approach where software techniques exploit state-of-the-art architectural features rather than relying upon the hardware-only development of lower power circuits and devices. Significant strides have been made in engineering lower power hardware components by reducing the supply voltage in CMOS circuits. This is illustrated in Fig. 1.2 that shows power consumption trends for microprocessors intended for use in mobile applications. Core voltages for these processors have decreased from 3.3 V in 1994 to under 1 V in 2004 while performance has grown significantly. However, hardware-only

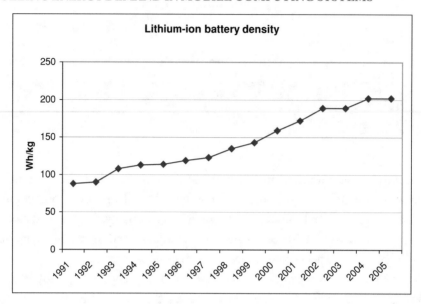

FIGURE 1.1: Rate of improvement of lithium-ion batteries [source of data: batteryuniversity.com].

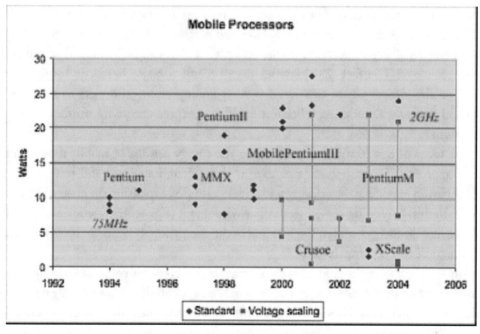

FIGURE 1.2: Trends in maximum power consumption for selected models of processor families aimed at the mobile market.

improvements are becoming more challenging to achieve. Effective power management will cut across the levels of system design. The processors offering the lower power values (shown as min–max ranges in pink) achieve it through dynamic voltage scaling (e.g., Intel's SpeedStep technology), a cross-level mechanism that can be exploited by the operating system for dynamic power management.

Fortunately, there are many opportunities to innovate on managing energy demand at the higher levels of a mobile system. Increasingly, device components offer low power modes that enable software to directly affect the energy consumption of the system. The challenge is to design resource management policies to effectively use these capabilities and to suggest improvements in those mechanisms to facilitate better techniques. Recently released processors can run at a range of different frequencies and voltages, raising the issue of how to schedule tasks to exploit these lower power operation points.

Mobile computing has been the initial motivation and incubator for much of the energy and power management research. As society recognizes the need for more energy conservation and the growing impact of computing technology on the demand for resources, innovations bred in the mobile computing domain eventually migrate into the mainstream of computing, providing both economic and environmental benefits. Adaptations for these techniques beyond the domain of mobile computing include reducing the noise caused by cooling fans in medical applications and reducing the electrical costs of operating and air conditioning large server installations.

1.1 LECTURE OVERVIEW

This lecture is aimed at new researchers and developers who are entering the area of power/energy management for mobile systems. In this section, we have motivated the need for explicit attention to energy management in mobile systems. The remaining chapters are organized as follows:

Chapter 2 provides the necessary foundations to work productively in this area. This discussion includes the definition of basic energy terminology and widely accepted metrics, the development of a system model of how power is consumed by the device, and the measurement methods and tools available for experimental evaluation.

Chapter 3 focuses on management of power state transitions for a single device. Idle time serves as the basis for policies that transition to lower power states. We explore the manifestations of idleness and the challenges involved in detecting idle periods. We describe examples of policies that, given an access pattern for the device, address the questions of when to power down to a lower power state and when to power back up to a higher power state. Then we consider techniques for modifying the access patterns of a workload to complement policies based upon using power state transitions.

Chapter 4 describes scheduling policies that exploit dynamic voltage scaling in processors. This leads to consideration of the quality of service (QoS) criteria that shape these policies.

Chapter 5 discusses the interactions and tradeoffs among the power management policies of multiple devices. We illustrate how the effective power management on one component of a system may have a negative impact on overall energy consumption. We introduce research efforts on whole-system energy management.

Chapter 6 explores the role of application-level involvement in energy management, ranging from providing minimal hints to the system-level power managers to energy-aware algorithm design. Several application program interfaces (APIs) have been proposed to provide information flow across the application-OS boundary.

Chapter 7 summarizes the key lessons of this lecture and discusses future directions in managing energy demand.

CHAPTER 2

System Energy Models and Metrics

This chapter presents basic energy terminology and widely accepted metrics. It describes mechanisms provided by hardware to allow systems to manage how power is consumed in mobile devices. It addresses the measurement techniques for calibrating system models and tools for experimental evaluation.

2.1 MODELS OF HOW POWER/ENERGY IS CONSUMED IN MOBILE DEVICES

In order to advance the state of the art in power/energy management, it is important to understand how power is currently used in mobile computing devices. This leads to building models of power consumption as a first step.

2.1.1 Power and Energy Models

One method of characterizing the power consumption of a computing platform is to identify its major hardware components and determine how much of the overall power budget each component requires. This is often presented in the form of a pie chart, as shown in Fig. 2.1. This kind of graph can be useful in focusing upon the major consumers within a device. However, it is important to understand how such a chart has been generated, what data have been included, and what assumptions have been made about the workload of the device. Unfortunately, such graphs often appear without being accompanied by this information that is crucial to correct interpretations. For example, Fig. 2.1 suggests that, after the CPU, the DVD is the next most significant power hog in this laptop computer. Depending on the goals of the project, this conclusion may not lead in a productive direction. For example, if the goal is to extend battery lifetime and the platform is to be used primarily for accessing email and the web, then the potentially high power consumption of the DVD is never seen in practice. This motivates our discussion of the uses of the terms *power* and *energy*.

This particular graph in Fig. 2.1 represents the maximum power consumption of each component in a Thinkpad R40 laptop, as described in Mahesri and Vardhan (2004). These data for the various components may be acquired from vendor data sheets or from

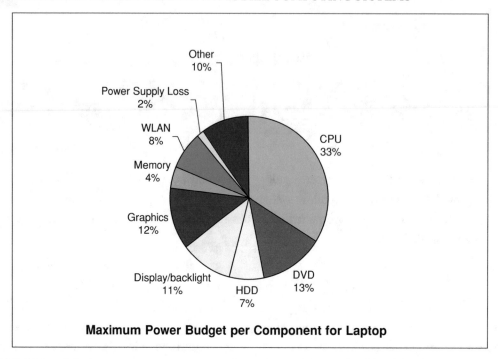

FIGURE 2.1: Maximum power measurements for each component of an IBM Thinkpad R40 Laptop: 1.3 GHz Pentium M, 256 MB, 40 GB HDD, 14.1 screen.

measurements using microbenchmarks that individually stress each single component. A synthetic microbenchmark to draw maximum power is sometimes referred to as a *power virus*. Vendor data sheets may report *thermal design power* or TDP rather than the true maximum power. TDP is typically defined as the highest sustained power that a real application can drive and is used for designing cooling solutions. We have composed these per-component peak power numbers into one chart to yield the breakdown in Fig. 2.1. The chart represents the worst-case power or highest rate of energy consumption that each component is capable of drawing. It can serve as a starting point in the absence of other knowledge about the workload or usage patterns for the device. It is probably impossible that any "real" program exists that can drive all components to their peak power simultaneously. However, this breakdown is useful if (1) the technical specifications are available, (2) there is no good information upon which to base assumptions about the intended or expected utilization of the device, and (3) the thermal limits of the device are of major interest.

Other profiles for this same platform use either energy or average power measurements by introducing workload assumptions and making the element of time explicit. Energy consumption is the focus when the limited energy capacity of the battery is the major concern.

The average power consumption of a laptop running a benchmark program is the total energy consumed during the execution of the benchmark divided by the execution time ($P = E/T$). Thus, power, when it is clear that the term is referring to average power consumption over a time interval, is often used interchangeably with energy consumption over the same time period.

One small step toward presenting a power profile based on average power is to scale the maximum power of each component by some estimated utilization factor to capture an assumed load (e.g., 60% of the time is spent actively using the CPU at peak power versus 40% at idle power). This is the approach used in Intel's Mobile Power Guidelines to estimate power consumption of future systems (Intel Corporation 2000).

In Fig. 2.2, the significance of the workload becomes obvious. The leftmost bar replicates the data in Fig. 2.1 in a stacked format expressed in Watts consumed by each component at its peak power. The remaining bars show average power consumption results from Mahesri and Vardhan (2004) with different benchmark programs. The 3DBench program (3D gaming) is generally accepted as a stress test for a machine. It achieves near peak power consumption by the CPU, display, graphics card, and memory. However, it does not exercise the WLAN,

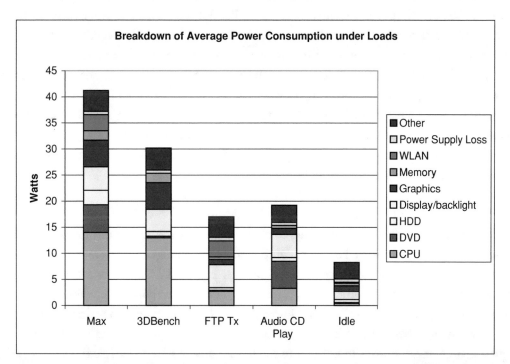

FIGURE 2.2: Average power breakdown for four benchmarks compared to the maximum per-component power breakdown of Fig. 2.1.

DVD, or disk. The other benchmarks represent a file transfer over wireless, playback of an audio CD, and an idle system. In each case, the power breakdown is significantly different, with the major energy-consuming components shifting from the CPU and graphics card in 3DBench to the display and WLAN for FTP transfer and the DVD and display for audio CD playback. The conclusion about which component will make the most important contribution to energy consumption for a target workload mix depends on how well its relative utilization of components matches one of these measured benchmarks.

The smartphone is an example that illustrates the impact of evolving usage patterns on the energy consumption estimates of a mobile device. Using the smartphone primarily as a traditional cell phone does not require constant backlighting of the display and avoids the relatively high power demand of that component. Estimates of battery lifetime by smartphone vendors are likely to be based on this traditional usage. However, other applications that are becoming popular (e.g., text messaging, taking photos) are likely to make the display a significant factor in determining the device's battery lifetime and a good target for power management attention.

Power models can be developed at other levels of abstraction. For example, characterizing the power costs of different instructions in the instruction set of the processor (e.g., Tiwari et al. 1996) can be useful in guiding energy-aware compiler optimizations to change the instructions generated in favor of lower power alternatives. Components at a lower level than considered in the breakdowns above are of interest in power-aware hardware design. Power models of components at the level of functional units and datapaths within a microprocessor architecture are relevant for clock gating and powering down blocks of circuitry. Higher level characterizations of an application domain may focus on events rather than physical components. For example, an environmental monitoring application of sensor networks may characterize the power costs of performing route discovery, reading a probe, sending a data packet toward the base station, and listening for messages from neighboring nodes.

2.1.2   Discrete Power States of Device Components

Models can be considerably richer than conveyed by either the maximum or average power consumption pie charts. Some hardware components offer a range of discrete power states that can be exploited in response to workload demands. Figure 2.3 illustrates a device with just two power states, drawing averages of p_{high} and p_{low} Watts in those states. When the device becomes idle, it can transition into the lower power state, incurring a transition cost in time and power which may spike if extra power is used to affect the state change. When a new request arrives, it can transition back up to the higher power state to service the request, again incurring a transition cost that can add latency before the request can be processed and even a spike in power as circuits power back up. Figure 2.3 represents this behavior as a piecewise

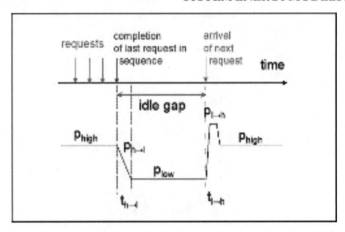

FIGURE 2.3: Timeline showing an on-demand power state transition.

linear approximation, not reflecting fine-grained power fluctuations. The values $p_{h \to l}$ and $p_{l \to h}$ represent the transitional power averaged over the transition periods $t_{h \to l}$ and $t_{l \to h}$, respectively. In the figure, $p_{h \to l}$ is shown as lower than p_{high} whereas $p_{l \to h}$ is higher.

The capability to transition between power states introduces the notion of the *breakeven time*. This is defined as the minimum amount of time that can be spent in and transitioning in and out of the low power state in order to make the transition beneficial in terms of energy consumption; thus, we denote it as $t_{benefit}$. The relevant parameters are shown in Fig. 2.4 where the benefit transitions are drawn in dashed blue lines and the transitions necessary to cover the idle time between the actual requests are shown in red. Since the transition times are assumed to be fixed, we just need to determine the minimum t_{low} such that

$$(t_{h \to l} * p_{h \to l}) + (t_{low} * p_{low}) + (t_{l \to h} * p_{l \to h}) < (t_{h \to l} + t_{low} + t_{l \to h}) * p_{high}.$$

Thus,

$$t_{low} = (t_{h \to l}(p_{h \to l} - p_{high}) + t_{l \to h}(p_{l \to h} - p_{high}))/(p_{high} - p_{low}).$$

Then we define the benefit as $t_{benefit} = t_{h \to l} + t_{low} + t_{l \to h}$. If the idle time, t_{idle}, is at least as long as $t_{benefit}$, then transitioning to the lower power state can save energy.

If we hope to avoid adding latency, we must assume that the beginning of the idle gap is recognized immediately with the transition down being immediately triggered and that the transition back up is triggered just in time for the end of the gap. In practice, policies that determine the triggering of transitions are based on observable behavior rather than assuming knowledge of forthcoming gap durations. For example, Fig. 2.3 shows the idle gap ending at an incoming request resulting in an on-demand transition that adds the latency of the transition to the performance. In addition, the policy that triggers a downward transition may spend extra

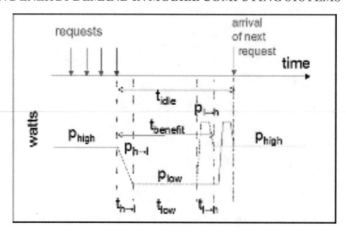

FIGURE 2.4: Breakeven time and idle time that justify a power state transition.

time in the high power state, drawing p_{high}, as when it uses a threshold of inactivity prior to classifying the device as idle. This overhead time must be deducted from t_{idle} before the $t_{benefit}$ comparison is performed in order to determine whether energy will, in fact, be saved. Policy considerations for making transitions are discussed in Chapter 3. We also discuss system-level policies such as caching (in Chapter 3) and application-level energy-aware adaptations (in Chapter 6) that modify request patterns to extend the idle periods and provide greater opportunity for the low-level mechanisms to exploit low power states.

Devices may have multiple power states rather than two as described above. The most familiar examples that are easily experienced by users include hard disks that stop spinning and displays that turn off when the system decides that they are not being used. Less noticeable examples include wireless radios and memory devices (DRAM and flash). Figure 2.5 shows multiple power states and transitions for a simplified hard drive. Spinning up and spinning down are transitions with significant costs (high power to spin up and large transition times, measured in the order of seconds), while the costs of other state transitions (narrow edges) are abstracted away in this model. Fully active states are those in which the disk is spinning and a read or write operation is in progress. Following the last queued request, this disk goes through a sequence of increasingly lower power states as more parts of the hardware are powered down (e.g., the disk may still be spinning but the heads are parked) on the way to the nonspinning standby state. Policies in the device firmware or operating system software determine what events will trigger each transition (e.g., thresholds of idle time) to exploit these multiple states.

Each type of device that offers multiple power modes has different parameters to capture in the model. The obvious issues to address are what are the set of states, how much power

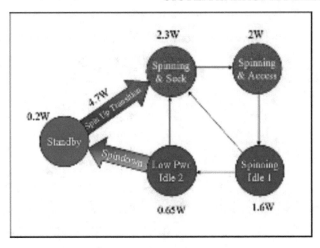

FIGURE 2.5: Power states of a generic hard disk drive.

is consumed in each state, and what are the transition latencies. It is also important to know whether the state transitions are amenable to software control through an established interface. For example, the Travelstar disk drives have an internal algorithm that controls the transitions among its idle states and do not provide an interface for external software control of them. The various power states also provide different functionalities. The disk must transition to a spinning state to service a request, but the interface is still active to accept an incoming disk request in a standby mode. A deeper sleep state than standby requires an explicit reset to reactivate. Displays may be still readable even when the backlight is turned off and the brightness of the backlight can be reduced. A radio for wireless networking can send messages at different transmission powers, affecting the range and noise. There is also the question of granularity of power control. In DRAM designed for mobile computing, each bank within the memory chip may be independently transitioned. In emerging display technologies (e.g., organic light emitting diodes), each pixel may be separately dimmed. Thus a model should capture the properties of each state including power consumption, transition latency, interface accessibility, granularity, and level of functionality.

As an example, Table 2.1 gives the power model for all components in all power modes for a Mica2 mote. The platform measured (Shnayder et al. 2004) consists of an Atmega 128 L processor, 128 KB of code memory, 4 KB of data memory, 512 KB EEPROM, and a ChipCon CC1000 radio. This illustrates three devices that exhibit multiple power states in different ways. The CPU has a choice of low power states (idle, ADC, power-down, power-save, standby, and extended standby) that can be entered via a SLEEP instruction with bit settings in a register to select mode. These are all nonoperational in the sense that the CPU cannot execute instructions

TABLE 2.1: Complete Power Model for Mica2 Mote (Shnayder et al. 2004)

MODE	CURRENT	MODE	CURRENT
CPU		Radio	
Active	8.0 mA	Rx	7.0 mA
Idle	3.2 mA	Tx (−20 dBm)	3.7 mA
ADC noise reduce	1.0 mA	Tx (−19 dBm)	5.2 mA
Power-down	103 μA	Tx (−15 dBm}	5.4 mA
Power-save	110 μA	Tx (−8 dBm)	6.5 mA
Standby	216 μA	Tx (−5 dBm)	7.1 mA
Extended standby	223 μA	Tx (0 dBm)	8.5 mA
Internal oscillator	0.93 mA	Tx (+4 dBm)	11.6 mA
LEDs	2.2 mA	Tx (+6 dBm)	13.S mA
Sensor board	0.7mA	Tx (+8 dBm)	17.4 mA
EEPROM access		Tx (+10 dBm)	21.5 mA
Read	6.2 mA		
Read time	565 μs		
Write	18.4 mA		
Write time	12.9 ms		

in any of these low power modes. The different modes affect the conditions required for wakeup of the CPU to resume execution. By contrast, there are multiple programmable transmission states for the radio. These are all operational states in that data can be transmitted in all of these states with not only different levels of power consumption but also different levels of delivered performance. Finally, the EEPROM shows different power consumption values that simply correspond to different operations being performed.

Power states are the key abstraction for device specifications in the industry-standard Advanced Configuration and Power Interface (ACPI) (Intel Corporation 2000). ACPI enables the operating system to control power management of compliant hardware. The specification defines global states for a system as a whole. These include the working state (G0), various sleep states (S1–S4 within G1) that capture familiar OS-specific states such as standby, sleep, hibernate, and safe-sleep, and power-off states (G2, G3) that necessitate rebooting to return to G0. It also defines device (D0–D3) and CPU states (C0–C3) with the possibility of operational lower power/performance states within the active states of D0 and C0. Figure 2.6 shows the ACPI states.

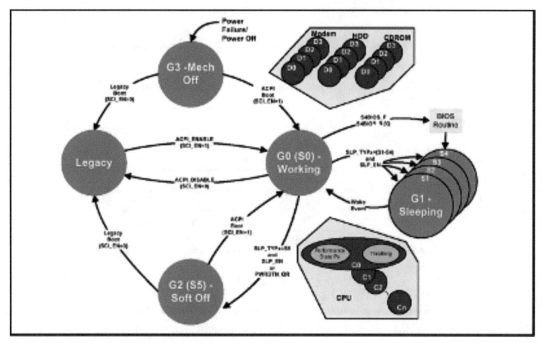

FIGURE 2.6: ACPI global states (Intel Corp. 2000).

2.1.3   Scaling Power Mechanisms

Another opportunity to exploit for energy savings arises from the properties of the CMOS digital circuits typically used in microprocessors. Power consumption in CMOS is composed of three factors: *dynamic* power consumption of switching logic, *short circuit* power, and *leakage* power. The major component and focus of most higher level power management is the dynamic switching power, which we denote as P_{SW}. $P_{SW} = A * V^2 * f * C$. In this equation, A is the switching activity factor, the probability of a gate switching state. V is the supply voltage. The clock frequency is f and C is the capacitance of the circuit. Reducing any of these terms can be effective in reducing P_{SW}.

Voltage scaling processors rely on scaling back the voltage, V, accompanied with necessary reductions in clock frequency, f. Voltage reduction is especially valuable because the power is related to the square of the voltage. However, the clock rate must also be lowered for the circuit to operate correctly because the settling time for a CMOS gate is proportional to the voltage. The lower the voltage, the longer it takes for the circuit to stabilize. There are limits in reducing the supply voltage; as supply voltage approaches the threshold voltage for the particular technology, circuit delay rises dramatically.

The relationship between V and f presents a tradeoff between performance and energy savings. If the workload demands are light and there exists idle time when running at the peak

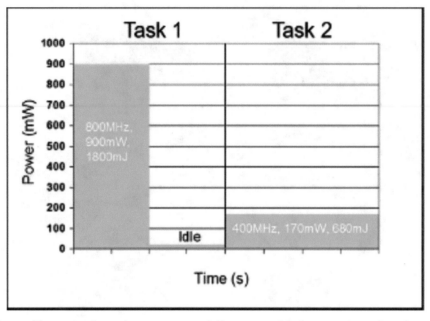

FIGURE 2.7: Two periodic tasks running at different speed/voltage settings (based on the Intel Xscale).

clock frequency, it is possible to reduce f and V without any impact on performance. This scenario is illustrated in Fig. 2.7 in which Task 1 running at a high clock frequency completes halfway through its period and then the processor becomes idle. Task 2 runs at half the speed and stretches its execution throughout the entire period. Because of the voltage scaling, the energy expended by Task 2 is significantly less than for Task 1. In other cases, degrading performance may be acceptable in order to achieve higher priority power goals (e.g., to avoid reaching a thermal limit). The policy choices for exploiting this capability (see Chapter 4) must determine when and to what levels to scale V and f.

Theoretically, one can view V and f as varying over a continuous range of values and offering the prospect of finding an optimal configuration. One popular abstract model is based upon the cubic-root rule that expresses power as $af^3 + b$ where the highest frequency at which the CPU can function correctly is proportional to the supply voltage. In practice, processors on the market provide a small number of discrete combinations of V and f across their ranges. Table 2.2 shows several examples of processors that offer voltage and frequency scaling. These processors may offer an interface for higher level software to affect a change. Transitioning between levels does incur a latency cost, influencing how often changing levels can be effective.

The idea of scaling across a range of power levels is not limited to processors. It has been suggested for displays and hard disk drives, as well. In the case of displays, this may take the

TABLE 2.2: Examples of Voltage Scaling Processors

PROCESSOR MODEL	VOLTAGE (V)	FREQUENCY (MHZ)
Intel 80200	1	333
	1.1	400
	1.3	600
	1.5	733
AMD Turion 64	0.9	800
	1.15	1600
	1.2	1800
Transmeta Crusoe 5900	0.875	433
	0.95	533
	1.05	667
	1.15	800
	1.25	900
	1.3	1000

form of gradually scaling the brightness of the backlight. Scaling the rotation speed of disks has also been proposed (Gurumurthi et al. 2003).

2.1.4 Simplified Battery Model

The final component in defining the energy-related system model for a mobile computing platform is the energy supply. As stated in Chapter 1, this lecture does not explore energy supply in detail and restricts its attention to available battery technology (lithium-ion, as the default). However, since extending *battery lifetime*, meaning the runtime from one full charge of the battery, is an important goal of energy management for mobile devices, we do need to establish some basic assumptions about the battery resources we aim to conserve, recognizing they simplify a complex topic.

The rated capacity of an n-volt battery is characterized in amp-hours (Ah). For example, one battery for the author's laptop is rated at 10.8 V, 4.4 Ah. The very simplest (and commonly used) model implicitly assumes that the voltage is constant and that the runtime is inversely proportional to the current drawn. This views a battery as a fixed pool of energy that is drained at the rate determined by power consumption.

In reality, there are many more parameters that could be included in a battery model, including chemistry type, operating temperature, age of the battery, number of discharge/charge cycles it has experienced, and discharge patterns. Voltage does not remain constant. In fact, voltage dropping below a specified "cutoff" threshold is what defines battery depletion. The voltage of a battery under steady load decreases with a slope that accelerates as it approaches this cutoff. However, the voltage level can temporarily recover with intermittent rest periods. During discharge, electrochemical reactions consume the active material near the electrodes. A rest period allows the chemicals in the cell to diffuse and restore the concentration of active material at the electrodes. This feature has motivated efforts to explore the benefits of bursty discharge patterns (Chiasserini and Rao 1999). In the case of lithium-ion batteries, the voltage is fairly stable until near the end where it drops rapidly. Next, the deliverable energy capacity is not really a fixed quantity but depends on the discharge rate. Ideally, a 10 Ah battery discharged at 5 A would last 2 h. However, at sufficiently high current loads, the effective capacity and runtime may be significantly less than expected due to losses from internal resistance. Thus, our 10 Ah battery discharged at 10 A might last only 30 min instead of the hour predicted by the linear model. The simple linear battery model breaks down for relatively high demands, but it can be reasonably applied for the lower range of discharge current.

Opportunities to incorporate energy supply information in power management projects exist because of the introduction of *smart battery* technology. A smart battery provides an interface to exchange information about the condition of the battery with the operating system. Such information may include the fully charged capacity, the present capacity, and the discharge rate. It may also provide runtime estimates and notifications of various levels of low battery. The interface also allows user setting of warning trip points. ACPI (Intel Corporation 2000) includes a standardized smart battery interface and compatibility specifications for battery manufacturers.

2.2   ENERGY METRICS

The choice of metrics to target for optimization reflects the goals of an energy management project. The development of the system energy model in the previous section has already described some of the features of the common metrics. Peak power (expressed in Watts) applies well to thermal problems. Average power (Watts) and energy (Joules) are often used interchangeably to measure effectiveness in conserving energy for a particular usage scenario or set of tasks. The workload assumptions are necessary since execution time is incorporated into these metrics. Total energy consumption for a workload is often used to estimate battery lifetime (hours), but the two are not perfectly related as shown in the discussion above of battery capacity under stressful loads.

There are other metrics that have been proposed for evaluating the effectiveness of power/energy management. Productivity metrics such as megabytes per Watt (MB/W), MFLOPS/W, or transactions per Watt make the work accomplished (the unit of work defined appropriately for the application) explicit in the metric. These are based upon the average power consumed in performing the work units of interest.

The metrics discussed so far do not directly address the energy/performance tradeoff. It is possible that improving energy consumption is being achieved at the expense of performance. Capturing this tradeoff is the justification for a single, combined metric such as *energy*delay* (usually specified without units). The advantage of this metric is that it imposes a penalty for either high energy consumption or high latency. While it is not the most intuitive metric, it serves a purpose by producing a single value that can be compared across many alternative design points. Another approach to capturing the performance tradeoff is to consider the energy metric, subject to quality of service (QoS) constraints. Examples include efforts to reduce energy such that no deadlines will be missed in a real-time system or so the latency of an operation will be bounded.

It is important to note the scope over which any of these energy metrics are being applied. It is common in energy/power management studies of a single component to report energy reductions for that component alone. For example, some of the earliest systems research on power management focused on spindown policies for hard disk drives (see Chapter 3) and reported simulation results on disk energy consumption. Saving energy for one component may have a little positive effect or even a negative impact upon the overall system energy consumption if higher power components become more active in response. These impacts may not be evident in measurement studies limited to one part of a system.

There are additional metrics that capture the global energy use of an ad-hoc network of mobile devices or a sensor network. For such environments, the battery lifetimes of individual nodes are important not just for local performance but also in how they contribute to the overall lifetime of the network. If just one node's battery dies, but it happens to be the only node that can forward messages to maintain connectivity between other nodes, then the network fails regardless of the average remaining battery capacity throughout the network. So, metrics such as time to network partition and variance in power consumption among the nodes are important in these applications.

2.3   MEASUREMENT AND TOOLS

In Section 2.1, we noted the need for measurement to calibrate system models of power/energy use and to help identify promising problems where energy might be saved. Methods for measuring or estimating energy use are also needed to experimentally evaluate proposed power management solutions. This section addresses the common challenges and methodologies at both stages of such a project.

2.3.1   Measurement Techniques

Designing experiments to measure power consumption in battery-powered mobile devices involves a number of issues including the accessibility of the platform under test, the measurement apparatus, the software to run during the tests, and the interpretation of the results. We begin with an elementary review of measuring DC current, tailored to our application.

The basic method employed for measuring the power consumption of a mobile platform is to connect a digital multimeter, reading current (denoted I, measured in amps, A), in series along the wire between the device and its power supply, as illustrated in Fig. 2.8. The alternative method is to measure the voltage drop across a resistor inserted in series with the power supply and then to calculate the current in the circuit using Ohm's Law ($I = V/R$). For a packaged, off-the-shelf platform, which is often the desired target for performing a power profile, this method yields the current drawn from the whole system in a fairly nonintrusive way. If the device

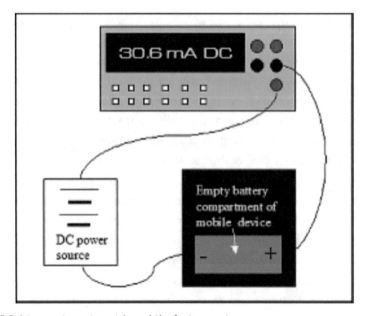

FIGURE 2.8: Multimeter in series with mobile device.

is battery powered, it is best to remove the battery and substitute an AC–DC adapter to provide a more stable power source. Then, the voltage can be assumed to be constant in calculating the power from current readings ($P = I * V$). Otherwise, relying on batteries during the test will introduce variability in the voltage supplied. Thus, with batteries as the power source, the supply voltage should be monitored throughout the test. In addition, if the battery is charging when the AC power supply is plugged into the device, the power measurements will reflect the recharging as well as the power consumption the test is intended to capture. The requirements of measurement apparatus for this level of study are modest. The desirable capabilities of a multimeter include the storage of sample values, flexible triggering mechanisms, high sampling rate, and an interface to a remote data collection machine.

An ideal situation is to have "self-contained" runtime energy estimation tools built into the device. At the present time, the instrumentation embedded in batteries and available via the smart battery interface of ACPI delivers inadequate detail for many measurement purposes. The reporting is coarse grained both in terms of data units and in terms of frequency of samples. There is work toward using processor hardware performance counters to track power-related events and estimate power consumption on the fly (Bellosa 2000, Contreras et al. 2005, Joseph and Martonosi 2001). Event counts are linked to a precalibrated power model to calculate power estimates. The available hardware performance counters provide information about processor and memory activity.

The method described thus far provides a single "external" point of measurement that gives the power consumption of the entire platform. A harder challenge is to derive the power consumption breakdown on a per-component basis. If one is fortunate enough to have access to development boards or platforms designed for experimentation involving power measurements, then probes can be inserted along the wires supplying power to individual components. Another possibility is to disassemble a packaged device to isolate the power supply wiring for those components that are not soldered directly onto the motherboard. Such an effort is described in detail in Mahesri and Vardhan (2004).

Without intrusive access to the internal wiring of the platform, per-component power consumption can still be found through indirect measurements. The indirect method of isolating the power consumption of individual components is based on a subtractive technique. The whole system is measured with the component of interest powered off or even removed, if possible, to serve as a baseline. Then synthetic software benchmarks are run to drive the component into all of its different power modes and the resulting measurements are subtracted from the baseline to give the power consumption attributable to that component in each state. Patterns artificially built into the current waveforms through the activity of the synthetic benchmark (e.g., a synch pulse of high current) can delimit the sequence of measurements that corresponds with the behavior of interest.

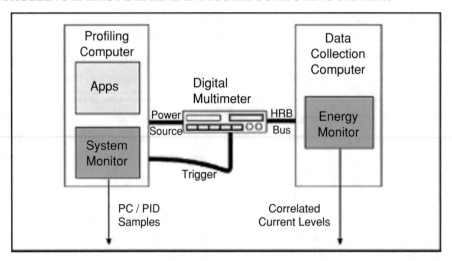

FIGURE 2.9: PowerScope (Flinn and Satyanarayanan 2004) © 2004 ACM, Inc. Reprinted by permission.

Measuring power consumption with a more realistic workload presents other challenges, especially in interpreting the results. The problem is how to associate the power consumption results with the corresponding computational behavior. PowerScope (Flinn and Satyanarayanan 1999a, 2004) is an example of an energy profiling tool that solves this problem of synchronizing traces of multimeter output and software profiling. The analysis phase attributes a percentage of total energy use to specific processes and procedures. The system is based on statistical sampling. The multimeter has an external trigger input and output and its clock controls the sampling. It first samples the power consumption of the system under test, the profiling computer in Fig. 2.9, and records the data on the data collection computer. Next, the multimeter toggles its external trigger line, causing an interrupt on the profiling computer so that the system monitor will record its execution state (program counter and process id). When that is done, the multimeter is triggered to enable another power sample. An off-line analysis tool associates each power sample with the process state samples and estimates energy usage in terms of constructs that are meaningful to the programmer.

2.3.2   Energy Estimation by Simulation
Given the obstacles in extracting fine-grained power measurements under realistic workloads instead of synthetic microbenchmarks, simulation is commonly used for evaluation. Coarse-grain measurements can be used to validate the overall energy estimates of the detailed simulators. There are two challenges in designing energy simulators: providing an accurate power model of the system and accurate timing of the simulated system behavior. Both are needed to deliver

energy results. Providing timing accuracy in a simulator has implications on its performance and scalability.

One approach has been to leverage execution-driven, cycle-accurate simulators from microarchitecture research and add power models to them. These include Wattch (Brooks et al. 2000) and SimplePower (Ye et al. 2000) for high-end architectures, XTREM for the Intel X-Scale (Contreras et al. 2004), and a simulator designed for embedded systems based on the StrongARM processor (Simunic et al. 1999). These simulators focus on the processor at the level of instructions or functional units, caches, and memory. The power models differ among these simulators, but are focused on capturing low-level capacitance or switching activity in the processor. The simulations based on these systems are accurate but slow.

The next level of abstraction incorporates more components of a complete computer system (e.g., I/O devices) into the simulator and enables monitoring of the execution of the operating system code as well as the application-level code. This motivates the development of full-system energy simulators, building upon existing full-system simulators such as SimOS (Rosenblum et al. 1997) and Simics (Magnusson et al. 2002). SimWattch (Chen et al. 2003) integrates Simics, running as an instruction-level functional simulator, with Wattch to study microarchitectures with the inclusion of operating system code. The functional simulator delivers simulation speed, handing off an instruction trace for cycle-level processor modeling to Wattch. The tricky issues involve carefully handling the synchronization of the two simulators to deal with speculation and exceptions. The power estimates have been restricted to those microarchitectural elements covered by Wattch, and I/O devices are not yet considered. SoftWatt (Gurumurthi et al. 2002) employs the multiple models in SimOS for modeling the processor, memory, and disk at different levels of detail (i.e., timing resolutions and power models). EMSIM (Tan et al. 2002) supports execution of Linux on an instruction set simulator of the StrongARM processor and simulation models of memory and other peripherals, each amended with Joule per cycle energy models. EMSIM has been validated against an Itsy evaluation board.

The speed and scalability are concerns for simulations at higher levels of abstraction, focusing on studies of I/O systems and networking. Again, the harder problem seems to be in dealing with time appropriately rather than in providing the system power models. File system research has depended on trace-driven studies using widely available trace data. Separating time-dependent and time-independent activities is key to using file traces in the Drive-Thru power simulator of the file storage hierarchy (Peek and Flinn 2005). This approach preserves the significant timings (e.g., disk idle time) that affect the results while accelerating the simulation speed compared to a simulation that accurately captures all timings. PowerTOSSIM (Shnayder et al. 2004) is a simulator for TinyOS-based sensor nodes that deals with the CPU timing issue by mapping basic blocks in the code executed by the simulation host machine to cycle counts in the corresponding code as compiled for the sensor mote. This allows accurate estimation of the time that would be spent in active versus idle (low power) modes by the simulated processor. Avoiding cycle counting during simulation runs results in lower overhead and enhances scalability, a desirable feature for simulating multiple nodes of a large sensor network.

CHAPTER 3

Management of Device Power States

This chapter begins by considering idleness and the challenges involved in detecting idle periods. We focus on managing power state transitions for a single device, exploiting the idle time offered by the access pattern for the device. We describe examples of policies that, given a workload, answer the questions of when to power down to a lower power state and when to power back up to a higher power state. Finally, we consider techniques for enhancing idle periods in order to improve the behavior of transition-based policies.

3.1 IDLENESS DEFINED AND DETECTED

At the top level, dynamic power management is based on powering down unused hardware until it is needed again. In Figs. 2.3 and 2.4 from Chapter 2, the idle time gap between the completion of an operation and the next request is used to describe the breakeven point for power state transitions to be beneficial. While the notion of idle time seems quite intuitive, some types of devices introduce nuances in how to detect or even define when they may be considered to be idle. Subtle differences in definitions may lead to novel opportunities to improve energy management strategies based on power state transitioning.

For some devices, the intuition about when they are idle is not captured well by the operational definitions that have been applied to them. One obvious example is the display. Typical display power management schemes are governed by the lack of keystrokes and mouse events, whereas the need for the display is determined by whether anyone is viewing the screen. The mismatch between the user input and the purpose of the display is illustrated by the familiar scenario of a display powering off during a lengthy discussion of a slide in a projected presentation. In one demonstration of redefining idle time to better match the purpose of the display, a low power camera is used to detect when no one is facing the screen (Dalton and Ellis 2003). In another example, there is no need to power the backlight when a sensor indicates that there is sufficient ambient light for a reflective screen to work well.

The processor is intuitively idle when there are no useful instructions to execute, but that can only be observed if the operating system does not fill the vacuum with idle processing or gratuitous maintenance daemons. In addition, the notion of "useful" work by application software is fuzzy. For example, is a busy-wait loop for synchronization a lost opportunity to exploit idleness of the processor? Just rethinking the standard definitions of when a device is considered to be idle may lead to innovation.

It is fairly straightforward to define when there are no pending requests for read or write operations directed to storage, ranging from cache to main memory to disks. However, this is an area that illustrates how policy decisions made at another level of the memory hierarchy can affect the requests seen at other levels. For example, changing the caching policy has an impact on the pattern of gaps in the request stream seen at a lower level of storage. The idle patterns are subject to change by interactions with different parts of a system.

For networking interfaces and devices, idleness is not purely a locally determined phenomenon. It is clear when there are no pending outgoing messages. However, whenever the radio is off and not connected to the network, it is not able to listen for attempts from the outside to communicate. This uncertainty about the possible arrival of incoming messages means that the device cannot know whether it can be considered truly idle. This has been a major complication in the power management of wireless radios.

Even when idle time is appropriately defined for the power management task at hand, detecting the beginning and predicting the end of a useful gap may not be simple, in practice. Different approaches to the problem of detecting idle periods have produced many of the variations on power transitioning policies in the literature. We discuss the range of approaches to this issue in the next section.

3.2   POLICIES FOR POWER STATE TRANSITIONS

There are a number of issues to address in designing a policy based upon power state transitions. These include if and when to transition into a lower power state, when to power back up to a higher state, what information to use in the policy decision, and how deep of a low power state to enter when there are multiple states from which to choose.

3.2.1   Transitions Among High and Low Power States

The typical power management scheme in products currently on the market is to provide a static threshold that may be either fixed or set by the user (usually specified in minutes) for components such as the disk, display, and system hibernation. Once the device has been idle (however that is defined for the component in question) for the threshold length of time, it enters a low power state. This is depicted in Fig. 3.1, which shows the delay imposed by the threshold before powering down and the on-demand transition back to the high power state. Note that this differs from both Figs. 2.3 and 2.4 in waiting through the threshold and, from Fig. 2.4, by not transitioning back up in anticipation of the next request.

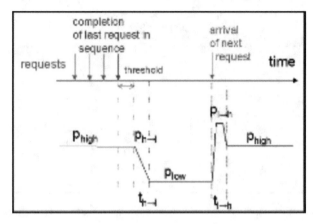

FIGURE 3.1: Threshold-based power state transition.

The underlying assumption of a threshold-based power transition policy is that observing an idle period that exceeds the threshold value predicts that the idle gap will continue for as long as the breakeven time; or the total $t_{idle} > t_{benefit} + t_{threshold}$. Given the power state model of a device and the distribution of idle gaps in a particular workload, it is feasible to choose a good timeout value for a static threshold-based policy for that workload. Such a study was done for hard disks with disk traces as a workload (Li et al. 1994) and found that the best threshold values were on the order of seconds rather than the longer thresholds typically implemented. Any such analysis is device and workload specific and any changes in the workload may call for a different threshold value.

Adaptive rather than fixed thresholds can address varying workloads. The decision to transition to the lower power state is still based on the idle time exceeding a threshold, but the timeout value currently in effect is a result of adapting to the previously seen access patterns. Many of the concepts related to dynamic management of device power states have been introduced for spindown policies for hard disks. For example, Douglis et al. (1995) propose an incremental threshold adjustment in reaction to the acceptability of the previous spindown–spinup session. Acceptability is defined in terms of the ratio between spinup delay and the idle time prior to the spinup. A spinup is deemed unacceptable, termed a *bump*, when the ratio exceeds a user-specified parameter value (e.g., 0.02–0.2). The threshold may be increased when a bump occurs and decreased when spinups fall into the acceptable range. Thresholds are adjusted using either additive (α_a, β_a) or multiplicative (α_m, β_m) parameter pairs with the α value determining the increase and β determining the decrease. For example, ($\alpha_a = 2$, $\beta_a = 1$) increases the threshold by adding 2s on a bump and decreases it by subtracting 1s on an acceptable spinup whereas ($\alpha_m = 1.5$, $\beta_m = 0.5$) multiplies the thresholds by the appropriate factor.

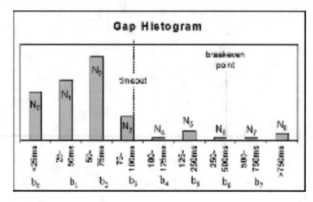

FIGURE 3.2: Histogram of gaps. Breakeven point calculated from device characteristics. Timeout is selected from a set of candidate thresholds corresponding with bucket boundaries.

The adaptive threshold selection has also been formulated in terms of a sequential rent-to-buy problem (Krishnan et al. 1999). The single rent-to-buy problem asks how long should one rent a resource before buying it when it is unknown how long it will be used. In power management, renting corresponds to remaining in the high power state for the threshold time, while buying corresponds to transitioning to the low power state when it is unknown how long the idle gap will be. The sequential rent-to-buy problem makes a series of such decisions where each new threshold value is informed by the history of previously observed gaps. The distribution of idle gap lengths is not assumed to be from a known distribution. The history of prior idle times can be captured in a histogram, carefully designed with respect to the number and sizes of buckets to ensure algorithmic efficiency.

Histograms of past idle times are widely used in dynamic power management (e.g. Anand et al. 2003). Consider the example in Fig. 3.2 that shows a possible gap distribution from a wireless network to illustrate how it can be used to determine an adaptive threshold. The number of gaps that fall into bucket i are denoted by N_i. The upper boundary of bucket i is b_i (e.g., $b_2 = 75$ ms). The breakeven time is based solely on device characteristics, namely, the power state model of an 802.11 wireless card given in Table 3.1. The transition energy cost, c, captures the latency and power costs of transitions down and immediately back up as $c = p_{h \to l} * t_{h \to l} + p_{l \to h} * t_{l \to h}$. The timeout is based on choosing the best candidate threshold from the set consisting of bucket boundaries, $b_{cand} \in \{b_0, \dots, b_m\}$, such that this timeout yields the lowest energy consumption for this particular access distribution. The energy consumption using threshold b_{cand} is calculated as follows:

$$E(b_{cand}) = p_{high} * \left(\sum_{i \le cand} (b_i * N_i) + \sum_{i > cand} (b_{cand} * N_i) \right) + c * \sum_{i > cand} N_i.$$

TABLE 3.1: Power State Model of Orinoco Silver Wireless Card (Anand et al. 2003)

STATE OR TRANSITION	POWER (W)	TIME (MS)
High power—listening	1.21	
Low power—idle	0.19	
Transition down (h → −l)	1.19	260
Transition up (l → −h)	1.04	230

An on-line machine learning technique has also been used to choose a timeout value as a function of the more recently observed disk activity (Helmbold et al. 1996). There remains a need for better solutions for aging out old data in order to quickly adapt to changing access patterns.

Rather than spend any amount of a timeout period in the high power state, single-value prediction-based methods attempt to predict the duration of the upcoming idle gap (Hwang and Wu 1997). When the gap arrives and its predicted duration is long enough, the device immediately transitions into the lower power state. The prediction of the gap length can also be used to transition back into the higher power, active state before the end of the gap and, ideally, just in time for the arrival of the next request, as depicted in Fig. 2.4. Whether or not this is an effective policy depends on the quality of the predictions that are made and the cost of being wrong in either energy expenditure or added latency. Exponential averaging of past gaps to estimate the next gap is the most commonly used technique. The overhead of performing on-line prediction has to be low compared to the requirements of the device it is used to manage. For example, prediction and decision-making for DRAM power state transitions must be much more lightweight than it needs to be for disk spindown.

There is also a body of theoretical work attempting to frame the power state transition decision as an optimization problem by modeling access patterns by known distributions (e.g., Simunic et al. 2000). Empirical studies of actual access patterns (e.g., Kotz and Essien 2005) tend to call into question the fit and stationary assumptions of these models.

3.2.2 Transitions Among Multiple Power States

When there are multiple low power states, there is the additional question of how to select which power state should be the destination for a transition. As discussed in Chapter 2, selectively and incrementally powering down the electronics in a device can lead to multiple power states,

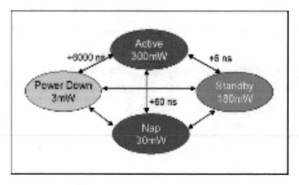

FIGURE 3.3: Power state model for RDRAM, showing power consumption in each state and transition times back to active mode.

each with different levels of power consumption, different transition costs, and different levels of degraded performance. An example of hardware with multiple states is DRAM designed for mobile devices (e.g., Rambus RDRAM, Intel Mobile-RAM) where the row and column decoders, sense amplifiers, and clock signal generation can be selectively disabled. Figure 3.3 gives the power state model for the RDRAM memory device. Figure 3.4 illustrates power transitions stepping through the full range of power states in the RDRAM chip, with thresholds at each stage determining the next transition (t_s is the threshold in the standby mode, t_n is the threshold in the nap mode). Servicing an access request is only possible in the active state, so the memory must transition back into the active mode at the incoming request. Policies do not necessarily step through all states. Developing a policy involves specifying which states to use and what threshold values to use in each power state transition.

There has been experimental exploration of appropriate threshold-based controller policies for such power-aware memories with multiple states (Lebeck et al. 2000, Delaluz et al.

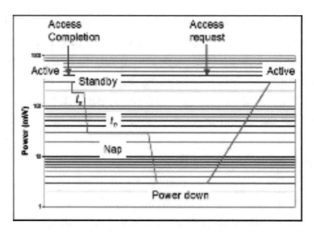

FIGURE 3.4: Stepwise power-down transitions for RDRAM.

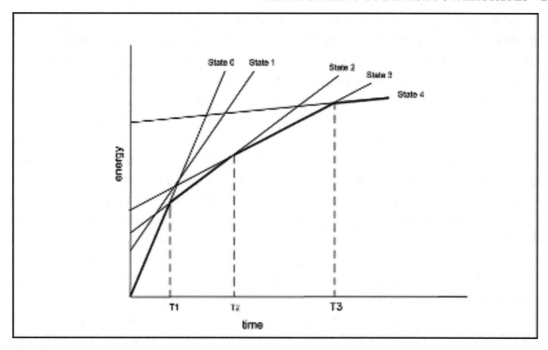

FIGURE 3.5: Energy consumption for a five-state system. For each line, the slope is the power used in that state and the y-intercept is the energy to transition up from that state. The thresholds between states are shown at the intersections of lines forming the lower envelope curve (Irani et al. 2003).

2001). For the architectures and workloads considered, these studies suggest that stepping down through all states sequentially is not the best choice, especially when caches act to filter the memory references, creating longer gaps. Delaluz et al. (2001) proposes a very simple hardware-based prediction scheme that estimates that the length of the next gap will be the same as the last gap and jumps directly to the power state deemed appropriate for that prediction, with results that show significant energy savings versus stepping through fixed thresholds. Taking a more theoretical approach, Irani et al. (2003) propose a competitive algorithm for the problem of determining a series of thresholds to step down through multiple low power states, requiring no information on the access patterns as well as a histogram-based method collecting data over a window of the last w idle periods. The algorithms throw out any power states that do not appear on a "lower envelope curve" such as that shown in Fig. 3.5. For example, State 1 does not contribute to the envelope, shown as a darker line.

3.2.3   Duty Cycling Modes

Regularly scheduled on–off duty cycling is an alternate way of exploiting the hardware power states instead of explicitly making each transition decision based on detecting idleness. Duty cycling the radio is the basis of the power-saving modes of wireless MAC protocols. The power-saving mode (PSM) in the IEEE 802.11 standard is built on top of the duty cycling of the radio, adding a protocol to provide coordination between endpoints to maintain connectivity. In PSM, a wired access point buffers data intended for the wireless network interface on the mobile device and regularly sends a beacon to indicate whether data are waiting. The client system periodically turns on its radio, on a static schedule, to listen for the beacon and goes back to sleep if no data are waiting. The typical beacon period is 100 msec. The radio is turned on if there are data to be sent from the mobile client. If the mobile client is receiving a transmission, its radio remains on for the duration of the transmission until it is complete. Otherwise, the wireless interface duty cycles at a fixed rate, regardless of communication patterns. The alternative to PSM is the continuously aware mode (CAM) in which the radio remains on and listening all the time.

Although there are significant energy-saving benefits to employing PSM, negative interactions affecting performance have been observed when communication patterns become constrained by the beacon timing. Adaptive policies have been proposed to vary the sleep-wakeup cycles in response to the traffic patterns. One policy resembling threshold-based transitions is to switch into CAM when the access point holds more than one packet for the mobile client and to switch from CAM to PSM after a timeout period with no packets received.

Self-tuning wireless power management (STPM) (Anand et al. 2003) adaptively switches coarse-grain power modes back and forth between PSM and CAM, basing its transition decisions on information about the transition costs (which can be significant because of the need to inform the access point), the base power consumption of the mobile computer, hints from the application on intended usage, and observed network access patterns. The use of application hints is proposed because the intermittent connectivity of network power management can interfere with automatically detecting true traffic patterns. One of the criteria for switching from PSM to CAM is the expectation, based on a hint, of a forthcoming transfer that is larger than a *breakeven size*. Another is the expectation, based on past access patterns, of a long enough run of individual transfers that together can justify being in CAM. Transitioning from CAM to PSM involves an idle time estimation and cost/benefit calculation. The histories of idle gaps and active runs are captured in histograms.

The bounded slowdown protocol (BSD) (Krashinsky and Balakrishnan 2002) increases the sleep time between wakeups that check for data buffered at the access point as a function of the increasing elapsed time since the mobile client's request in request-response traffic patterns. This essentially defines a TCP connection as idle, from the mobile client's perspective, following an awake period with no response to a client request. The device then begins intermittent listening with the growing cycle time. BSD bounds the increase in RTT by a protocol parameter p.

In the realm of mobile, ad-hoc and sensor networks, there has been significant effort in developing energy-aware protocols to ensure connectivity while allowing the radio to be cycled on and off. These protocols generally establish a schedule for neighboring nodes to turn on their radios, listen for and exchange data, and then turn them off again to save battery resources. Nodes may also disable their radios throughout the remainder of transmissions they have overheard, but which are not intended for them (e.g. PAMAS (Singh and Raghavendra 1998)). Establishing a schedule involves the solution to other problems such as time synchronization, route discovery to determine neighbors, and mutual agreement on the schedule.

3.2.4   Granularity of Transitions

Our discussion of power state transitions has focused primarily on timing. Some technologies also offer opportunities to take advantage of spatial granularity of power management. In power-aware DRAM, the ability to fully exploit the available power states benefits from having portions of the memory to which power transitions can be independently applied and that can be distinguished from each other based on activity levels (see Section 3.3.2).

Another example involves emerging display technology. The idea of zoned backlighting, proposed in Flinn and Satyanarayanan (1999b), would partition the screen into zones that could be selectively illuminated and content could be adapted to exploit the zones effectively (e.g., reducing the size or moving images to span fewer zones). In displays made of organic light emitting diodes (OLEDs), the power consumption of each pixel is independent and related to its brightness and color. Color may be viewed as a form of an implicit power state in contrast to illumination. This fine-grained control offers opportunities to adapt the energy consumption to different usage patterns incurred by different applications and user preferences. In Iyer et al. (2003), a characterization of typical display usage was performed in a user study involving a range of test users on laptop and desktop machines. The results show that, on average, only about 60% of the screen area is used by the "window of user focus". Results show differences among users and among application workloads in terms of display needs for space and for qualitative features such as range of colors (e.g., Photoshop has more stringent requirements than pop-up alert messages). The proposed solution is called Dark Windows and it selectively changes the brightness and color of screen areas outside the active window. Modifications to the background screen areas that are considered include partially dimming, fully dimming, changing to gray scale, and changing to the target OLED technology's most power-efficient green color as depicted in Fig. 3.6.

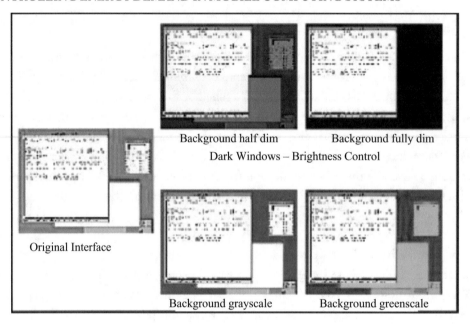

FIGURE 3.6: Dark Windows (Iyer et al. 2003).

3.3 MODIFYING REQUEST PATTERNS TO INCREASE IDLE PERIODS

Given a power management scheme that will react to idle time by manipulating the power states of the device, there exist opportunities for higher level systems to reshape the patterns of those idle periods to make the power state transition policy more effective. We present several examples that demonstrate this theme.

3.3.1 Caching and Prefetching

Caching is a fundamental systems technique to improve performance. It is applied at various levels of the storage hierarchy, by copying actively used data into a higher level of memory. The cache filters out requests from the request stream that can be served from the faster cache and reduces the number of accesses that pass through to the slower lower level storage. This effect naturally increases the interarrival times of requests that are handled by the lower level and creates idle opportunities that may be exploited by dynamic power management. However, the design of the cache has typically not been optimized for the sake of effective power state transitions and any energy benefit is usually just a side effect of the primary performance goals. The implementation details of the caching layer may even work against power management. For example, the typical periodicity of the OS daemon that writes back dirty disk blocks from the disk cache is frequent enough to reduce the risk of losing updates but can also frustrate threshold-based disk spindown.

With an increasing emphasis on energy consumption, the redesign of caching policies to facilitate power management is drawing attention. We discuss this issue in the context of the operating system's file buffer cache and disk subsystem. The problem is to design policies for the buffer cache that shape the disk request stream in order to create beneficial blocks of idle time for spinning down the disk and to amortize power state transitions across operations. Efforts in this area (Weissel et al. 2002, Zeng et al. 2003, Papathanasiou and Scott 2004) have addressed many aspects, including application interfaces. For now, we consider the techniques for making the buffer cache management more compatible with disk power management within the operating system without using any knowledge about the applications (cf., Chapter 6).

Deferred writes are one of the ways in which the cache filters the disk request stream. Multiple write operations to the same file block are applied to the cached copy and not written back to disk immediately. Thus, a sequence of file system writes can be absorbed into a smaller number of disk operations (ideally one), with a window of vulnerability when the dirty cached block and block on disk are inconsistent. The delay of the write-back tries to balance the risk of losing data from volatile memory and the ability to coalesce as many individual file writes as possible into a single disk write. For example, the default settings for the timing of the Linux update daemon (pdupdate) specify that it run every 5 s and flush dirty buffers that are older than 30 s. Unfortunately, a modern disk drive is likely to have a breakeven time that is greater than 5 s. This introduces a third factor, the disk spindown characteristics, into the balancing act required in the writeback policies. Among the changes that have been proposed is simply increasing the allowable age of a dirty buffer that forces a spinup for writeback. For example, there is now a "laptop mode" for Linux that can be configured to increase the maximum lifetime of dirty buffers in cache whenever the machine is running on battery. This weighs the tradeoff between power consumption and risk of data loss in favor of energy savings. It is also useful to be more opportunistic about an already spinning disk by writing back all dirty data before the disk spins down, even if a dirty block is officially too young and is likely to still be accumulating file writes. This affects the tradeoff between power and the number of disk requests, potentially generating rewrites for a dirty block that was still actively being written. While having explicit information about an impending spindown (e.g., Advanced Configuration and Power Interface) may be convenient, these deferred writes only need to tag onto other disk operations that have triggered a spinning up of the disk. Figure 3.7 illustrates the effect of disk-friendly modifications of the update function. The power consumption between spikes of activity and the overall energy use are significantly reduced with energy-aware caching.

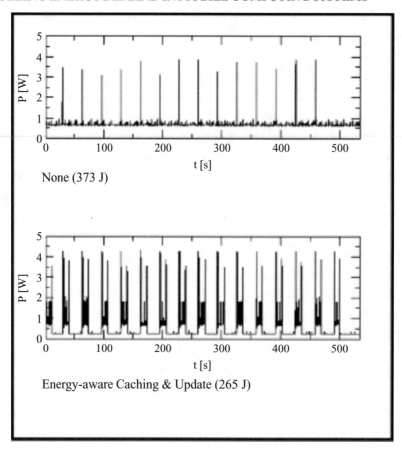

FIGURE 3.7: Default Linux writeback activity (top) and energy-aware update daemon (below). (Weissel et al. 2002) Reprinted by permission.

The cache also filters out file read operations that reuse a block between the time a disk operation fetches the block into a buffer and a replacement operation evicts it. File prefetching and replacement policies influence the remaining cache misses. The traditional designs for these policies have focused on performance goals, and revisiting these techniques to encourage better idle times for power management is warranted.

Prefetching for energy involves solutions to the problems of identifying which blocks to prefetch, determining how many such blocks are needed at a time, allocating buffers for them, and deciding when to initiate the disk operations for prefetched data. Aside from simple, but commonly used, sequential access patterns, information from the application via hints can be valuable in selecting which blocks to prefetch (Papathanasiou and Scott 2004) and we return to this topic of using application-specific knowledge in Chapter 6. It is instructive to relate the problem of prefetching for energy to hoarding for disconnected operation in mobile file systems (Kistler and Satyanarayanan 1991, Kuenning and Popeck 1997). In both cases, the objective is to provide enough cached data for the application to use while the file system is unavailable—whether it is because of disconnection or to enable long idle periods for disk spindown. This argues for aggressive prefetching when the disk is spinning to satisfy the need for data during long idle gaps, assuming free buffers can be acquired. This bursty activity is one difference from traditional, performance-related prefetching where data can be smoothly transferred into the buffer cache with the goal of a block arriving prior to the first request for it. In Papathanasiou and Scott (2004), the authors argue that it is better to speculatively fetch many data blocks when the disk is active, even including some data that the applications may never use if doing so can help avoid on-demand read requests that require the disk to spin up. They propose an epoch-based system with an active phase for prefetching and flushing dirty buffers and an idle phase. The idle phase ends when predictions indicate that it is time to initiate a new active phase to avoid a miss, a demand miss actually occurs, or the system needs memory resources.

Even if each application process can prefetch the data, it needs to survive long periods without access to the disk, the disk may not get the opportunity to spin down unless the disk I/O of all the processes can be roughly synchronized. One role that buffer allocation can play is to globally coordinate the I/O needs of running processes so that they all have run out of cached data or have filled their empty buffers at approximately the same time. A solution is to allocate the number of buffers to a process based on the process's rate of data consumption and data generation in the buffer cache (Zeng et al. 2003, Papathanasiou and Scott 2004).

An energy-aware replacement algorithm can also influence the disk idle periods by changing the capacity or eviction misses that appear in the disk request stream. Most work on this topic targets multiple disk systems in data centers rather than the mobile computing environment. However, there is a discussion of the issues in the context of a single disk off-line replacement algorithm in Zhu and Zhou (2005). The idea is to evict the block that has the lowest energy penalty to refetch if and when it is requested in the future. Intuitively, if the evicted block is requested close in time to a cold miss that requires the disk to be active, then refetching it has less of an energy cost than if it triggered the disk spinup by itself. The tradeoff is that this replacement strategy does not minimize the number of misses, but rather the energy cost of the misses. There is as yet no on-line realization of this idea.

Traffic shaping by a server proxy that is communicating with a wireless device is another example that can be viewed as a buffering solution designed to create more usable or predictable idle periods for the client to make effective use of its radio power states. For example, in Chandra and Vahdat (2002), popular formats for multimedia streaming are considered and the limitations of client-side policies to predict gaps in packet arrivals are discussed. Consequently, the authors propose an architecture to shape multimedia packet transmissions at a proxy within

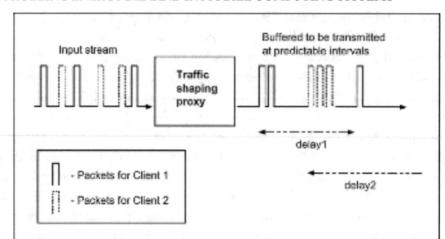

FIGURE 3.8: Policies to shape traffic in order for packets to arrive at clients at predictable intervals (Chandra and Vahdat 2002).

the network infrastructure—either at the media server or close to the wireless access point. Figure 3.8 illustrates the effect of a traffic shaping proxy transmitting media packets with appropriate per-client delays to enable radio energy savings without causing performance loss. One of the caveats of this technique is the need to counteract a media player's adaptations to packet reception delays.

3.3.2 Memory Management

In Section 3.2.2, we focused on the memory access gap distribution experienced by power-aware DRAM devices. However, the overall pattern does not necessarily affect every memory module the same because of finer grain control, as suggested in Section 3.2.4. The operating system can play a role in modifying the memory access patterns directed to independently power-managed memory nodes. We use the term "node" to indicate the smallest granularity over which power state transitions can be applied (e.g., a chip or a bank). Our first example illustrates a collaborative approach between the OS and the memory controller. Power-aware page allocation (Lebeck et al. 2000) tailors the content of memory nodes to complement the lower level power state management. A sequential first-touch page placement policy is effective in grouping virtual pages of an application into a set of physical pages on a node so that they are transitioned together. Pages with similar activity characteristics are clustered in a small footprint instead of being scattered across several memory nodes that might otherwise be sufficiently idle to go into a lower power state. This involvement of the OS greatly improves the *energy*delay* results over those achieved with the hardware power state transitions alone. In

Chapter 6, we describe compiler optimizations that rearrange data structures of an application to accomplish a similar goal within the user code.

Power-aware virtual memory (PAVM) (Huang et al. 2003) assigns the responsibility of making the transition decisions solely to the OS and also employs virtual to physical page placement decisions. The idea is to allocate the pages mapped into a process's address space to a preferred set of memory nodes so that the active set of pages has a small footprint. When a context switch occurs, the active nodes of the next process to run are transitioned to the standby power state and those of the previous process are transitioned down to the nap mode. The transition latency can be hidden by the other overhead of the context switch. This is implemented in the Linux kernel with a NUMA-like memory management layer to do the preferential physical page allocation. The initial experiences with the system revealed how some of the traditional OS optimizations for space and time, namely shared dynamically-loaded libraries (DLLs) and the file buffer cache for reuse of file blocks, interact with the new goal of compact process footprints for power management. One lesson is that DLLs must be

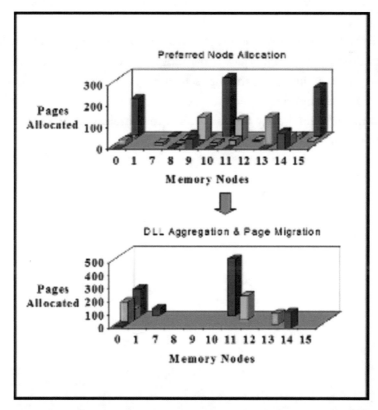

FIGURE 3.9: Power-aware virtual memory page placement (source of data: Huang et al. (2003)).

handled separately to prevent those pages from being pulled into the various preferred sets of the processes sharing them and being spread all across memory. The solution proposed is to use sequential first-touch allocation with DLL pages. The second lesson concerns the scattering impact of file reuse through the buffer cache. The proposed solution is to migrate pages to better placements in preferred nodes. This is accomplished by a migration daemon and incurs additional overhead that must be kept limited (e.g., avoiding migration for short-lived processes). Figure 3.9 shows the distribution of pages across memory nodes for four processes under the original allocation and the more compact per-process footprints with the policy enhancements. Experimental results show significant energy reductions with both the basic system and the improvements. A similar system is discussed in Delaluz et al. (2002).

3.4   SUMMARY

In this chapter, we have shown that the idle periods in the request patterns for a device that are offered by the workload are what allow policies to be developed that can exploit the low power states available in the hardware. Threshold-based policies are common, with the timeout values as the key parameters. Adaptive threshold selection can respond to observations of the recent history of idle gaps.

The proactive manipulation of access patterns is a powerful approach to make power state transitioning more effective. Techniques that change the temporal and spatial characteristics of the accesses include caching, prefetching, scheduling, and placement. The basic idea is to lengthen idle gaps and amortize transition overhead among a cluster of requests. The goal is essentially to manufacture bursty access patterns to complement power state management policies that detect idleness.

CHAPTER 4

Dynamic Voltage Scheduling (DVS)

This chapter discusses scheduling policies that exploit dynamic frequency and voltage scaling in processors. The strength of scaling frequency and voltage together is that it provides quadratic energy savings with only a linear performance cost, as explained in Section 2.1.3. Table 2.2 gives several examples of processor chips with voltage scaling capabilities. In current practice, these features are used in simple policies such as setting voltage and frequency based on whether the power source is AC or battery. However, the potential for fine-grained policies that exploit dynamic voltage scaling lies in using the slowdown to squeeze out processor idle cycles that might occur at the end of a task when it is run at the maximum processor speed, replacing those idle cycles with continuous processing at a lower speed and voltage level.

There is an interesting contrast between the goal of continuous processor activity and the bursty usage behavior that benefits devices which transition into nonoperational low power modes considered in Chapter 3. This tension between the smoothing behavior of dynamic voltage scheduling (DVS) and the bursty behavior for power state transitions is especially important when it affects the interactions among multiple components that use these different capabilities in their energy management. Changing the speed of the processor will have an impact on the generation of requests for other hardware resources. Such interactions, both positive and negative, are considered in Chapter 5.

In comparison to the detection of and reaction to idle time in managing low power states, the scheduling of frequency and voltage changes is *inherently predictive*. Once idleness has been observed, it is too late to extend the immediately preceding work into that idle space. The predictions must provide enough information to guide the major DVS decisions about *when* to adjust frequency and voltage as well as to what *settings* they should be changed.

The presentation begins by considering the quality of service (QoS) criteria that formulate what is an acceptable tradeoff between energy consumption and performance and the impact of workload assumptions on defining that tradeoff. We then consider the impact of the system model on the solution space. We discuss the major categories of DVS solutions and issues involved in combining them into a working system.

4.1   WORKLOAD AND QUALITY OF SERVICE CRITERIA

The simplest case for DVS is depicted in Fig. 2.7 from Chapter 2 with two instances of a periodic task that has a static processing demand. In the second execution of the task, a lower clock frequency/voltage combination can fill the period with active processing, resulting in energy savings. This example is oversimplified in a number of ways (e.g., the trivial match between the fixed load and the frequency settings); however, even this simple case exposes the fundamental question of DVS: *When and how is it acceptable to delay the completion time of a task for the sake of saving energy?* The assumption in this example is that the periodic behavior corresponds to deadlines and finishing a task before a deadline has no particular value to the application. In this case, DVS can save energy without a significant impact on the performance goals.

The DVS problem is often formulated as an optimization of energy consumption, subject to performance constraints that can justify slowing down the processor. Assumptions about what is known about the workload determine what constraints are appropriate. It becomes a different problem depending on whether the workload is categorized as hard real time, soft real time, interactive, or a general-purpose mix. If it is a real-time workload, then is the worst-case execution time (WCET) known deterministically or probabilistically? Are there multiple programs that are time sharing the system and what are their priorities? These workload distinctions, to a large degree, provide the organization of this chapter. Interval-based solutions (Section 4.3) assume practically no a priori knowledge about the workload and target general-purpose workloads that do not have well-defined deadlines. The performance constraints are generally based on *utilization* of the processor falling into a desired range or relative to the maximum speed setting. Real-time workloads (Section 4.4) offer much more information including WCET, periods, and deadlines. The QoS constraints are based on *missed deadlines* and the degree to which missing deadlines may be acceptable (e.g., no missed deadlines or a bound on the fraction of deadlines missed). Complications arise when the processing demand of a task is variable rather than fixed so that even an optimal speed-setting solution based on WCET does not preclude idle time remaining prior to the deadline. In non-real-time workloads, deadlines can still be exploited as a justification for degraded performance if they can be automatically derived (Section 4.5). For example, acceptable performance degradation for interactive processes can be defined in terms of delays of user interface events (serving as deadlines) that may be perceived by users.

4.1.1   About Time and Idle Time

The frequency scaling aspect of DVS affects the definition and monitoring of the QoS performance constraints. Characterizations of the work performed by tasks need to be expressed in a frequency-independent metric (e.g., cycle counts or execution time at the processor's highest speed setting). Thus, it is better to talk about worst-case execution cycles (WCEC) rather than WCET that depends on the frequency setting in effect. Deadlines, periods, and monitoring intervals are expressed as real time.

Implementation implications of DVS on timing involve knowing how "time" is reported by the system. Different hardware provides different mechanisms and the developer needs to be aware of whether the timer data accessed depend on the current speed or are invariant to frequency changes. Some timers even stop advancing whenever the processor is in a sleep mode which may hinder the measurement of processor idle time.

Figure 2.7 can also be reinterpreted to illustrate an important distinction between "hard" and "soft" idle times, as articulated in Weiser (1994). Not all idle time is equally amenable to being squeezed out by slower processing. Suppose the idle time shown in Task 1 actually represents waiting for the delivery of data in response to a disk read request that was generated at the end of Task 1, and Task 2 is simply the processing of that data on its arrival. If Task 1 were slowed down, it would only delay making the I/O request and do nothing to reduce the disk latency. Soft idle time involves waiting for an independently initiated event, such as the start of a new period for real-time tasks or the arrival of an external request. Low processor utilization is not necessarily indicative of a great DVS opportunity.

4.2   SYSTEM MODELS OF FREQUENCY/VOLTAGE SCALING

Another factor that distinguishes DVS problem statements is where they aim on the spectrum between an ideal system model and actual hardware and, if a solution initially assumes an abstract model, whether there is a mapping into discrete settings. The continuous model only makes sense when the quality of load predictions is high enough to support seeking an optimal speed/voltage setting, favoring those types of workloads with more information available. Figure 4.1 shows a least square fitting of the form $a * f^3 + b$ for the discrete settings for the Intel XScale processor, a particularly close fit for commercial processors.

Consider an example in which it has been determined that a periodic task's optimal speed setting (calculated as its WCEC/period) is 700 MHz, a choice that is not available on the XScale. One question is how to map that desired setting into the discrete settings offered by the hardware. Selecting a single hardware setting for the task implies rounding up to 800 MHz to avoid missing a deadline, but that will use more energy than necessary. If we allow one adjustment point during execution of the task, the execution time can be evenly split between 600 MHz and 800 MHz or evenly split between 400 MHz and 1000 MHz. This is another oversimplified example that illustrates several points. There are two alternatives presented here that average to the desired frequency. The best choice for switching at the midpoint of the period is the combination that will minimize the energy consumption. This is the pair of immediately adjacent settings to the optimal value. In this example, pairing 600

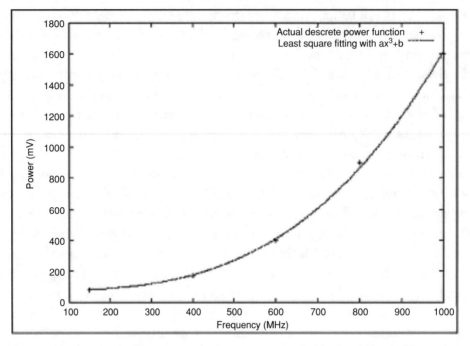

FIGURE 4.1: Abstract power function versus discrete settings for the Intel Xscale (Xu et al. 2004) ©
2004 ACM, Inc. Reprinted by permission.

and 800 MHz consumes an average of 650 mW whereas pairing 400 and 1000 MHz consumes
an average of 885 mW. If the optimal setting is based on the WCEC and the task is likely
to complete earlier than its worst case, then the ordering of the two discrete settings matters.
It is better to use the slower speed first and, hopefully, finish before executing for very long
(or at all) at the higher speed. In this example, the timing of a single transition is assumed to
be halfway through the task execution. In general, the number and timing of transitions is a
nontrivial problem that yields various solutions. This discussion illustrates a common approach
in the literature of, first, solving the problem for a continuous power model and, then, mapping
the solution onto the hardware. The alternative is to solve the problem directly for the realistic
discrete model. The nuances of this issue are discussed further in Section 4.4.1.

Other issues of the system model include whether the costs of making speed and voltage
changes are considered and whether the idle processor state is accurately captured. Transition
latencies on current voltage scaling processors can be significant (e.g., 75–150 µs for AMD processors), limiting how often a practical solution can afford to make such changes. Formulating
the problem as one of determining the optimal setting for each cycle of a task's execution leads
to solutions that cannot be implemented in the current technology, but may apply if transition
costs are reduced in the future.

Much of the research in DVS considers only the energy consumption of the processor. One aspect of the system model that matters in the effectiveness of DVS for the whole system is how frequency scaling affects other components that are not scalable. In particular, scaling the processor speed when the speed of external memory does not scale changes the relative memory latency. In addition, there are potential interactions between DVS and the request patterns to other components that trigger dynamic power management based on idle gaps (Chapter 5).

4.3   INTERVAL-BASED APPROACHES

Interval-based DVS algorithms assume no information about the workload or modifications to user applications. The problem is defined as scaling the frequency and voltage in order to reduce energy consumption while keeping performance, typically measured as processor utilization, within acceptable bounds. The general approach in interval-based schemes is to break execution into fixed length intervals and to use the utilizations measured in past intervals to predict the utilization for the next interval and adjust the frequency (and corresponding voltage) to bring the performance toward the desired range. Solutions in the literature differ in several ways: (1) how the performance goal is precisely defined, (2) how much of the history is used for prediction, (3) whether performance is monitored on a system-wide or a per-task basis, and (4) how and when the processor clock is changed.

The most commonly used metric in interval-based DVS is processor utilization, the percentage of nonidle time during the interval. This may be captured in terms of active cycle counts or busy times recorded at each clock interrupt. Thresholds on utilization define the desired range of performance. Users may provide input on what is acceptable. Early research (Weiser 1994) used a measure of the amount of work (excess cycles) carried over from a previous interval when the processor ran too slowly during an interval to finish it; however, such information is not available for practical implementations. The interval length is important in balancing responsiveness to changes and the ability to observe longer-term patterns. Most studies that propose an operating system-level solution assume an interval in the 5–100 ms range.

An interesting alternative (Childers et al. 2000) uses an interval-based approach at the architectural level (with 2 μs interval) to adapt to variations in instruction-level parallelism (ILP). The metric is a target rate for MIPS that can be achieved by increasing the clock speed during phases of lower ILP (e.g., branch instructions) and scaling frequency back when high ILP can produce the desired MIPS rate. A hardware counter giving the number of committed instructions determines the observed MIPS rate.

The classic interval-based DVS algorithms first predict the utilization for the next interval and then change the clock speed if the predicted value is above or below the thresholds. The simplest algorithm is to use the utilization of the previous interval to predict the next one, called

PAST. Another common algorithm is an exponential moving average of previous intervals with a weight parameter. This is often referred to as AVG_N where N is the weight. The weighted utilization at time t, $W_t = (N * W_{t-1} + U_{t-1})/(N + 1)$.

Based on the prediction, the speed settings may be changed. Grunwald et al. (2000) considered three policies: *One Step*, which increases or decreases the clock frequency to the next discrete speed level, *Peg*, which sets it to the highest or lowest speed, and *Double*, which doubles or halves the speed. Weiser (1994) increased or decreased the frequency by a percentage of the maximum speed. Alternatively, a particular setting can be calculated directly based on a prediction of the required work in the forthcoming interval.

Figure 4.2 shows a simplified example of the standard predictors using the *One-Step* speed adjustment. The workload is shown as the utilization under the maximum frequency and all idle times are assumed to be soft idle times. This establishes how many active cycles of work are required by this workload. The starting frequency is assumed to be 600 MHz for the two predictors, PAST and AVG_2. The thresholds are 95%, signifying that a prediction greater than that triggers an increase in speed to the next discrete level and 85% such that a prediction lower than that triggers a decrease, with the goal of raising utilization. In this example, 1000 MHz is the top speed and 400 MHz is the lowest speed considered. For PAST, the prediction in the table below each interval in the graph is the utilization from the previous interval. In response, PAST starts by decreasing speed because the prediction is lower than 85% and then, when the utilization reaches 100%, it triggers successive speed increases until the demand drops. The prediction table for AVG_2 gives the weighted utilizations with $N = 2$ for each interval. AVG_2 slowly incorporates the increased utilization but the predictions remain below the lower threshold until interval 6, with the speed staying at the lowest level. Finally, the predictions cross the upper threshold at interval 9 and cause speed increases. By this time, the workload demand has increased as well, keeping utilization high.

Other prediction algorithms are possible and several have been proposed with promising simulation-based results. However, experimental evaluations of interval-based DVS based on implementation (Grunwald et al. 2000) have made a case that this approach suffers because of the limited workload information available. It is difficult, in general, to strike a balance between responsiveness to changes in the workload, incurring the overhead of many frequency transitions, and sluggish reaction to processor demand when longer history is incorporated. Irregular workloads are a particular challenge. Intervals may be useful in conjunction with other approaches (Section 4.5). For instance, the Vertigo system (Flautner and Mudge 2002) offers a hierarchy of speed-setting policies, with each level using different workload characteristics. At the bottom of their set of complementary choices, there is a "perspectives-based" interval scheduler, with adjustable history windows capturing utilization on a per-task basis. If higher

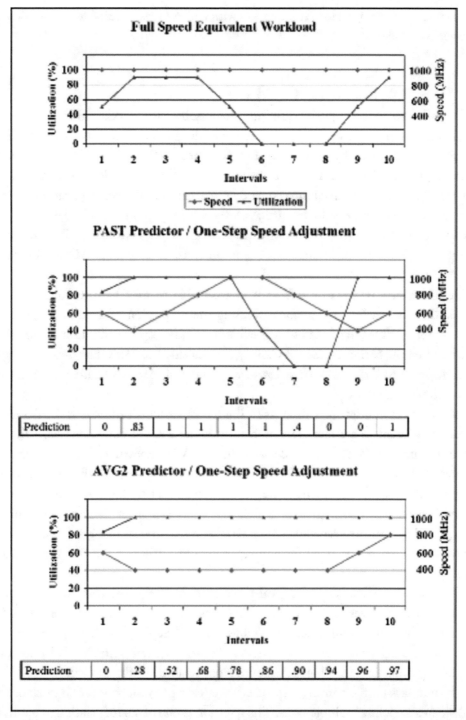

FIGURE 4.2: Classic interval predictors and one-step speed adjustment. Thresholds 85% and 95%.

level policies have better workload information with which to make decisions, they will override this interval scheduler.

4.4 DVS FOR REAL-TIME TASKS

Real-time task systems offer a great deal of information about the workload that DVS algorithms can exploit and, thus, have been the focus of a wealth of research in this area. A single real-time task is characterized by its release time, R, deadline D, and WCEC. The worst-case execution time is the WCEC divided by the maximum frequency: $WCET = WCEC/f_{max}$. Periodic tasks have regular arrival times such that release times of the task instances correspond with their deadlines and periods. In particular, the ith invocation, T_i, of a periodic task with period P occurs at $R_i = (i - 1) * P$ with deadline $D_i = i * P$. Since an instance may not require its WCEC, the actual number of execution cycles is denoted as AC and its execution time is AC/f.

Deadlines may be hard or soft. It is unacceptable to miss hard deadlines. With soft deadlines, there is usually a requirement limiting the proportion of deadlines missed or the severity of miss. WCEC and AC also provide a terminology for idle time. The idle time that exists when the period exceeds $WCEC/f_{max}$ is called *static slack*. The difference between $WCEC/f_{max}$ and AC/f_{max} gives us *dynamic slack*. The goal of real-time DVS solutions can be viewed as eliminating slack.

In systems of multiple real-time tasks, the scheduler must prioritize task invocations so that deadline guarantees are met. The problem assumes a set of periodic tasks, $\{T_1, \ldots, T_n\}$, with the jth invocation of task T_i denoted by T_{ij}. Well-known real-time schedulers include Earliest Deadline First (EDF) and Rate Monotonic (RM). EDF bases priorities on which task has the most imminent deadline. RM assigns priorities based on the periods of the tasks. Schedulability tests guarantee that a task set can meet its deadlines. For example, a task set has a feasible schedule under EDF if $\Sigma_{1 \leq j \leq n} WCET_i/P_i \leq 1$. Both EDF and RM have been used as the basis for adding frequency setting to the real-time scheduling decision.

There are two major classes of real-time DVS research: intratask and intertask. The intratask problem involves adjusting clock frequencies during the execution of a task. There are several motivations for intratask speed adjustments including simply mapping an optimal solution for reducing static slack based on WCEC onto discrete settings as discussed earlier and exploiting dynamic slack when actual executions require less than WCEC.

The intertask problem involves speed changes on a per-task basis (at dispatch or context switch) when scheduling multiple tasks. The question is one of sharing slack that accumulates as tasks finish early with other tasks scheduled to run in the future so that they will have more resources to finish by their deadlines and can afford to slow down. This focuses on redistribution of slack among tasks.

4.4.1   Intratask DVS

The question addressed by intratask approaches is how to deal with dynamic slack within the tasks that cause it by not always requiring the specified worst-case execution. One approach is to model the distribution of actual task demands and schedule a speed schedule based on the probability of slack. Another approach is to monitor the progress being made by a task execution to detect slack and do a midcourse correction in the speed settings. Both approaches involve the problem of determining the number and placement of transition points during execution where the frequency and voltage adjustments may be performed.

The first approach recognizes that the actual execution cycles of tasks are often less than the WCEC. AC is a random variable with a cumulative distribution function, $F(x)$, that gives the probability that AC $\leq x$. For hard real-time tasks, $F(\text{WCEC}) = 1$. The problem is to devise a speed schedule that reflects the distribution of AC in order to avoid the creation of slack. This involves first deriving the distribution. The Processor Acceleration to Conserve Energy (PACE) project (Lorch and Smith 2001) derives a theoretical formula based on the task's $F(x)$ that expresses the speed schedule as a continuous function of time. Applying this would require that the speed of the processor is capable of increasing at every cycle of execution which is not possible in real machines. Thus, this schedule must be approximated with a piecewise constant schedule with a limited number and granularity of transition points. PACE proposes using quantiles of the distribution as transition points. The distribution of task work requirements is not usually known a priori and must be estimated by observing the history of similar task executions. Various sampling methods are explored in Lorch and Smith (2001) and the distributions are estimated from the samples either by assuming a known family of distributions (e.g., gamma) and estimating its parameters or by using a nonparametric method.

A similar project, GRACE (Yuan and Nahrstedt 2003), uses histograms to capture the AC distribution and bin boundaries to serve as the transition points. We illustrate the technique with an example from GRACE since it is somewhat simpler to describe. In Fig. 4.3, we present an example of a cumulative distribution of cycle demand and its estimation by a histogram of r bins with the count in bin i being the number of task instances that use between b_{i-1} and b_i cycles. Figure 4.4 shows a speed schedule and how it affects three different tasks. The speed schedule has a scaling point every 10^6 cycles at which point it changes to the designated speed. Thus job_1 which uses only 1.6×10^6 cycles experiences only one scaling point at cycle 10^6 whereas job_3 which needs 3.9×10^6 cycles passes through all three scaling points during its execution.

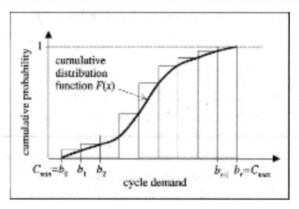

FIGURE 4.3: CDF of cycle demand and histogram bins (Yuan and Nahrstedt 2003) © 2003 ACM, Inc. Reprinted by permission.

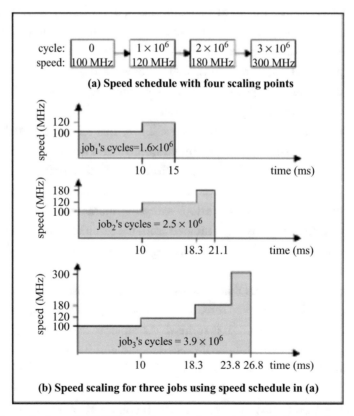

FIGURE 4.4: Example from GRACE showing speed schedule and three tasks with different cycle demands (Yuan and Nahrstedt 2003) © 2003 ACM, Inc. Reprinted by permission.

Computing the piecewise constant schedule using the bin boundaries $\{b_0, b_1, \ldots, b_m\}$ as transition points, with s_i as the size of the ith bin (i.e., $s_0 = b_0$; $s_i = b_i - b_{i-1}, 0 < i \leq m$), involves solving the following optimization problem:

$$\text{minimize} \sum_{0 \leq i \leq m} s_i(1 - F(b_i)) f_{bi}^2$$

$$\text{subject to} \sum_{0 \leq i \leq m} (s_i * 1/f_{bi}) \leq P.$$

The goal is to find a speed for each bin boundary, f_{bi}, that minimizes the energy consumption while satisfying the execution time constraint given by the task's period, P. The speed for the s_i cycles between subsequent bin boundaries is uniform.

Practical PACE (PPACE) (Xu et al. 2004) also follows the probabilistic approach, but assumes a more realistic power model than PACE and GRACE. PPACE provides a polynomial time approximation algorithm that accounts for the nonzero power costs for an idle processor, charges for making transitions, and directly targets the discrete frequency/voltage levels supported by real processors.

The second intratask DVS approach is based on monitoring the progress of a task during execution and detecting, early enough to respond with speed changes, whether the task is on track to finish well before its deadline. The problem is often articulated as redistributing the dynamic slack created during a task's execution to the remaining processing by the same task, thus enabling a frequency and voltage reduction. Redistributing slack may also go in the other direction, essentially borrowing unrealized slack from the predicted future execution path to start slowly and pay it back later in the task's execution with higher frequencies if misprediction threatens the task by missing its deadline. The problem becomes how to evaluate a task's *progress relative to its worst case* during execution to uncover potentially useful dynamic slack. What runtime information can be found to suggest that the AC for the current task will be less than WCEC? This generally requires knowledge about the program, often provided in the form of annotations (hints or explicit actions) inserted by the compiler.

To illustrate this approach, we present an operating system solution that uses relatively lightweight compiler hints to convey relative progress (AbouGhazaleh et al. 2006). It breaks the task into segments such that the speed and voltage can be reconsidered for each segment. The granularity of segments affects the overhead of scaling speed and voltage. The OS controls this overhead by determining the interval between periodic interrupts that trigger speed adaptation. These interrupts that serve as opportunities for speed and voltage scaling are called power management points (PMPs). In handling a PMP, the OS reads a memory location (WCR) that holds the most recently written value for the worst-case remaining cycles. This value has been calculated by an instrumentation code called a power management hint (PMH) and

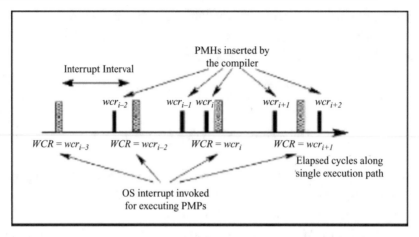

FIGURE 4.5: Power management hints, PMH (inserted by the compiler) and power management points, PMP (invoked by OS) (AbouGhazaleh et al. 2006) © 2006 ACM, Inc. Reprinted by permission.

reflects the progress of the task. The OS uses WCR and the remaining time until the deadline to compute a new speed. There are two schemes, one that redistributes any slack proportionally to all remaining segments and one that greedily allocates it to the next segment. The proportional formula is $f_{next} = $ WCR/(D − current time − transition overhead).

Prior to execution, the compiler inserts PMHs throughout the program code to update the WCR value based on the path of execution being followed. For example, if the executing task takes a branch that bypasses a long path, the WCR exposes the resulting dynamic slack. The PMHs are spaced to ensure that at least one PMH runs before the OS takes each PMP interrupt. The placement of PMHs in the code is guided by offline profiling for the cycle counts of code regions and the structure of the program control flow. The PMHs just leave the WCR value for the OS to use in PMPs rather than directly causing scaling actions.

Figure 4.5 shows the PMHs (five points denoted by short dark bars) encountered along a particular execution path for the task. These are spaced to occur between OS interrupts for PMPs (denoted by hashed bars at regular intervals). When the OS takes a PMP interrupt, it reads the latest value of WCR. This value reflects progress at the most recently executed PMH, but not the task's progress at the actual time of the interrupt. For example, at the third PMP interrupt, the OS reads the value WCR = wcr_i and it never sees the value wcr_{i-1} written earlier in that interrupt interval.

4.4.2   Intertask DVS

The problem of intertask DVS is to add the dimension of speed scheduling to the real-time scheduling for a task set of periodic tasks. The goal is to save energy without missing deadlines.

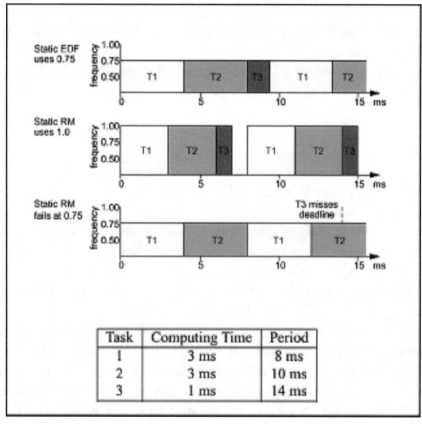

Task	Computing Time	Period
1	3 ms	8 ms
2	3 ms	10 ms
3	1 ms	14 ms

FIGURE 4.6: Task set and scaling for WCET under EDF and RM (Pillai and Shin 2001) © 2001 ACM, Inc. Reprinted by permission.

The frequency adjustments are performed when tasks complete or are dispatched rather than during execution. Most work is based on modifying the EDF or RM scheduling algorithms. Figure 4.6 gives an example of a task set scaled for EDF and RM schedulers.

The first step is to consider scaling when all tasks require their WCEC. For the system model of continuous speed adjustments, the problem is to find a constant scaling factor to be applied to all tasks that will not violate the QoS constraints. It is assumed that the lowest speed for which the task set is schedulable yields the lowest energy consumption. Recall that we define WCET as WCEC/f_{max}. Let α $(0 < \alpha \leq 1)$ represent a scaling factor such that the optimal frequency is $f_{opt} = \alpha * f_{max}$. For EDF, the problem becomes finding the smallest α such that the task set is schedulable according to the modified test:

$$\sum_{1 \leq j \leq n} \text{WCET}_i / P_i \leq \alpha.$$

Thus, for the task set in Fig. 4.6, the schedulability test yields $3/8 + 3/10 + 1/14 \leq 0.75$. In this example, the frequency is rounded up to one of three discrete settings, with scaling factors of 0.5, 0.75, and 1.

The schedulability test for RM is similarly modified with the scaling factor as follows: for all T_i in the task set ordered by period length, $\{T_1, \ldots, T_n | P_1 \leq \ldots \leq P_n\}$,

$$\sum_{1 \leq j \leq i} (\lceil P_i / P_j \rceil * \text{WCET}_j) \leq \alpha * P_i.$$

This accounts for the number of instances of T_j that will precede the execution of T_i. Thus, for Task 3 in our task set and a scaling factor of 0.75, the test fails with $13 > 0.75 \times 14$.

The next step is to consider dynamic slack in scheduling and frequency scaling of the task set. In the intertask DVS problem, the dynamic slack is shared among the task set such that if a task finishes before using its WCEC, then the slack can benefit those tasks following it in the schedule up to some point. There are a rich variety of techniques for estimating and redistributing dynamic slack. For a greedy solution such as the dynamic reclaiming algorithm (DRA) (Aydin et al. 2004), the beneficiary is the next task. For the cycle-conserving algorithms (Pillai and Shin 2001), the set is rescaled as if the WCET of the task that just finished early was actually $AC/f_{\max}$ until the task's next release when the scaling computation reverts to using its original WCET. We describe DRA and the cycle-conserving version based on EDF (ccEDF).

In ccEDF (Pillai and Shin 2001), the scaling computation is performed on each task release and task completion. The computation is based on the static scaling for EDF, rewritten as $\Sigma_{1 \leq j \leq n} U_i / P_i \leq f_i / f_{\max}$, where f_i is the lowest of the available discrete frequencies that satisfies the inequality. U_i are set to WCET_i / P_i on each new release of task T_i and set to $AC_i / (f_{\max} * P_i)$ on task T_i's completion. Then the scaling is recomputed based on the updated U values. The dynamic slack created by instance j of task i, T_{ij}, is essentially spread across a window from completion of T_{ij} to release of T_{ij+1}. Figure 4.7 shows ccEDF applied to the task set of Fig. 4.6, but with the actual execution times instead of the worst case.

DRA (Aydin et al. 2004) is also based on EDF. The idea behind DRA is to track the actual execution relative to a canonical schedule S^{can} that is the scaled schedule based on WCEC. Whereas early task completion in S^{can} represents dynamic slack, the benefit available to the next task is represented by the *earliness* of its dispatch relative to S^{can}. This is not necessarily the entire amount of dynamic slack if, say, the next task had not yet been released when the previous task completed early. Thus, the earliness is the quantity of unused computation that can be applied toward lowering the frequency for the next task. The actual schedule is never allowed to fall behind the canonical schedule in this scheme. The problem then becomes how to estimate earliness efficiently. The algorithm is based on a data structure that models the

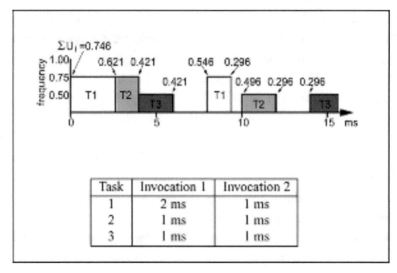

FIGURE 4.7: Example of cycle-conserving EDF (Pillai and Shin 2001) © 2001 ACM, Inc. Reprinted by permission.

progress under S^{can} for comparison with the actual progress and is updated on task arrivals and completions.

There are also solutions that are much more aggressive about anticipating that dynamic slack will arise, scaling optimistically, but ensuring deadlines are met by reserving enough full speed capacity to handle the worst case. These schemes do get ahead of the canonical schedule but can recover at high speed. For example, the look-ahead algorithm based on EDF (Pillai and Shin 2001) plans in reverse EDF order to defer work as late as possible in the schedule (betting that it will not be needed), thus reducing the number of cycles that must be allocated early and can be done at lower frequency.

4.5   TOWARD THE GENERAL-PURPOSE ENVIRONMENT

While the earliest work on DVS focused on interval schedulers and no knowledge about the applications, the information available for a real-time workload has been attractive and effective in developing DVS solutions. Deadlines have been a convenient rationale for performance slowdown. In order to move toward DVS schemes that can support a more general-purpose mix of applications, some researchers have proposed automatically inferring deadlines in non-real-time processes based on interactive behavior or communication patterns (Lorch and Smith 2001, Flautner et al. 2001).

The first category of behavior that appears amenable to automatic deadlines is the interactive episode triggered by a GUI event such as pressing the keyboard or mouse. The associated task set is tracked through the chain of communicating processes initiated by the GUI event. The end of the episode is defined as when all of these tasks are done and their data consumed. The deadline for an interactive episode is defined as the perception threshold that is typically considered to be around 50 ms. There is evidence from human–computer interaction (HCI) research indicating that interactive users will not detect performance degradation, say from lowering the clock frequency, below this perception threshold.

Other OS-observable communication patterns may suggest episodes with implicit deadlines that can be exploited for frequency scaling. Periodic episodes show little variation in run length. Producer–consumer episodes define the producer's deadline based on when the consumer needs its data. The producer can be slowed down to produce that data just in time. These techniques involve tracking on a per-process basis.

Since there are different solutions targeted to different classes of application, a system designed to support a general-purpose environment must either choose the lowest common denominator solution and sacrifice any extra knowledge that may be offered by an application or else it must find a way to mix and match a set of scheduling algorithms. The latter is the approach proposed in Vertigo (Flautner and Mudge 2002). Vertigo offers a mechanism called the policy stack for installing different performance-setting policies and an interface to specify how to blend their independent decisions. The implementation described in Flautner and Mudge (2002) demonstrated the perspectives-based interval scheduler at the bottom of the stack and one using deadlines based on UI events at the top. The middle layer provides a place for specialized policies to be developed and deployed in the future. The authors argue that the multiple policies compensate for their individual weaknesses. The question of how such an architecture for incorporating multiple interacting policies will actually function seems to be open for further exploration and there are some current efforts in this direction (Gurun and Krints 2005).

4.6   SUMMARY

In this chapter, we have described various approaches used to dynamically schedule the frequency and voltage settings for the processor. The majority of effort in this area has focused on real-time workloads that provide information such as deadline requirements that can be used to define slack times and justify slowing down the execution. Variations on DVS techniques include intratask algorithms that make scaling decisions during execution of a task and intertask algorithms that distribute dynamic slack among a real-time task set. The importance of knowledge about the workload for effective DVS is emphasized in the attempts to automatically infer deadlines in non-real-time application domains such as interactive episodes. There appears to be a tension between the DVS strategy of running "slow and steady" and the bursty patterns of behavior favored for devices employing low power modes. This is an issue we consider in the next chapter.

CHAPTER 5

Multiple Devices—Interactions and Tradeoffs

While the majority of work in energy management for mobile computing has focused on a single component at a time and the savings in consumption by that device, this chapter discusses the interactions and tradeoffs among the power management policies of multiple devices. There are both negative and positive implications in considering multiple devices. We illustrate how the power management on one component of a system may have a negative impact on overall energy consumption or motivate policy changes for another device. Opportunities exist in systems that make choices among alternative devices on the basis of energy consumption (e.g., remote versus local computation, storage on flash versus disk). Finally, we introduce research efforts that focus on whole-system energy management.

5.1 IMPACT OF DEVICE ENERGY MANAGEMENT ON OTHER COMPONENTS

In reporting energy savings of a voltage scaling processor under DVS, it is often acknowledged that stretching out execution time may require other subsystems to consume more energy by remaining in their active states longer. Changing processor frequency can also change the timing of I/O requests generated by the running program. Slowing the speed does not address idle cycles that are caused by blocking I/O. These types of interactions can obviously dilute the benefits of managing one component in isolation.

In particular, the interaction between memory and voltage-scaled processors is one area that has been explored (Martin and Siewiorek 2001, Grunwald et al. 2000, Fan et al. 2003). These interactions can complicate the simple system model that assumes performance is proportional to frequency. In that model, the lowest frequency and voltage combination is assumed to always deliver the most energy savings. This leads to speed-setting solutions with the goal of determining the lowest speed that satisfies the quality of service (QoS) requirements. However, memory speed does not scale along with the processor. Memory bandwidth becomes relatively more important at the higher end of the processor's frequency range. For applications with high cache miss rates, memory accesses may limit the ability of frequency increases to deliver the expected performance for meeting a deadline. For example, consider an intratask dynamic voltage scheduling (DVS) solution (e.g., the speed schedule in GRACE (Yuan and Nahrstedt 2003) as presented in Fig. 4.4) that starts at a low speed and speeds up at scaling points as task execution continues. The high speed settings at the latter scaling points may not be able to make up for the slow start if memory limitations arise to throttle performance expectations.

If the memory technology on the system can go into a low power state during the slack time of a periodic task, then memory energy costs may dominate processor energy savings at the lowest frequency settings where that slack is eliminated. The "sweet spot" for energy savings of the DVS processor/memory combination may be at an intermediate frequency that has enough slack to allow the memory to use its lower power mode while still reducing processor energy relative to its maximum frequency/voltage level. Figure 5.1 shows this effect. This figure adopts a variable voltage processor model based on Intel XScale and memory technology based on Mobile RAM (275 mW in active mode, 75 mW for standby, and 1.75 mW for power-down). The "naïve" power-managed memory controller policy puts the memory into power-down during slack times. The total lines in the figure show the energy consumed by memory added to CPU energy. The best point for the energy consumed by CPU with naïve memory is not the lowest frequency setting but rather 200 MHz, violating the simple model. The "aggressive" memory controller uses a threshold-based policy to transition into lower power states for fine-

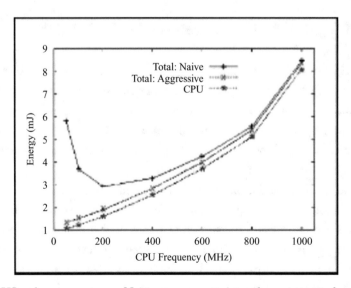

FIGURE 5.1: DVS and memory energy. Naive memory goes into a low power mode when the processor is idle. Aggressive memory does fine-grain transitions during execution. Total refers to memory + CPU energy.

grain access pattern gaps throughout task execution. In this case, memory energy consumption remains a small contribution to the total over the entire frequency range.

Interactions have an impact on the design of power management policies. As discussed above, the speed-setting algorithm must take memory bandwidth into account. On the other hand, DVS has an impact on the distribution of idle time gaps that may be used in determining thresholds for power state transition policies as described in the example illustrated by Fig. 3.2. At different frequency settings, the same workload characteristics produce different gap histograms and, therefore, different optimal threshold values for transitioning to the lower power state. Thus, the interaction with DVS adds another factor to the dynamic power management policies of other system components such as network and I/O devices.

5.2   ENERGY-AWARE ALTERNATIVES

One of the potential benefits arising from jointly considering the energy management of multiple devices on a system is the ability to "play off" one device against another. If there are several different ways to achieve a particular functionality, the system can choose the one that saves the most energy based on conditions at the moment. In this section, we consider examples involving alternatives for computation (e.g., local versus remote execution), storage (e.g., disk versus flash), and networking (e.g., 802.11 versus Bluetooth).

5.2.1   Computation Versus Communication

One example of trying to exploit a multiple device tradeoff is dynamically determining whether a computational task on a battery-powered wireless platform should be computed locally or transferred to a remote server for processing. From the point of view of the mobile computer, this is a tradeoff between using its battery power for the local processor or using power for the radio to communicate with the server. Presumably, the resources on the mobile device can sleep while the remote server is busy with the computation.

Whether or not remote execution can actually save energy on the mobile device depends on a number of factors. The first consideration is whether the task demands enough computation to justify sending the work off the platform. This may involve assessing whether the computation is well suited to the local processor (e.g., whether there is support for floating point operations). The second issue is how much communication is required to find an accessible compute server, send the request, send the input data, and retrieve the final results. This gives rise to a range of assumptions about where the code for the computations and where the data files reside. For example, if it is assumed that the mobile computer is already participating in a distributed file system, then the communication cost involves maintaining consistency of updated file copies. However, the network file system allows the mobile device to avoid shipping entire files for each remote execution request since they would already be available to the server. The final consideration is how effectively the mobile platform can exploit a low power state while waiting for the response and how well it can predict when it is the right time to wake up to get the results. Deciding whether a particular task at a particular time can benefit from remote execution is a challenging problem.

System frameworks have been proposed to support this dynamic local versus remote execution tradeoff (Rudenko et al. 1999, Flinn et al. 2002). Resource monitoring and prediction are the key differences in these systems. The common features include a distributed file system with support for consistency with file replicas on the mobile platform as well as on the server infrastructure. This removes some of the on-demand data transfer that might be associated with a remote process invocation. There is also a code library of programs that have been registered and installed on the server as remotely executable services. This avoids code migration for every request and provides a database for storing per-process execution history. The gathering of history and development of models for predicting future demand are areas in which solutions differ—from occasional relearning the power costs of a process to developing models that track the task's use of several resources. The decision making is based on this learned history of demand. Spectra (Flinn et al. 2002) incorporates current resource availability of local and remote CPU load, local and remote file cache state, and network characteristics into its decision making and selects the option that maximizes the user's utility function.

5.2.2 Storage Alternatives

Local versus remote file storage is another example of a tradeoff, but the design space becomes even richer when multiple local storage devices are possible, each with their own properties and power state models. For example, a mobile platform may support a hard disk drive, flash memory, and various removable memory devices (e.g., USB memory, microdrive card) in addition to the wireless network that provides access to remote storage. The most energy-efficient option for storing data may depend on the present power state of devices (e.g., whether the disk is already spinning or requires a spinup). Exploiting the heterogeneity of storage choices in the design of a file system for the mobile computer involves reinterpreting the notion of a storage hierarchy. The hierarchy is not static when storage devices have low power states, when network variability affects access to remote storage, and when removable storage devices temporarily disappear from a working system and even reappear elsewhere. This requires a more adaptive system model with online monitoring of the current state.

These are some of the issues addressed by the Blue File System (Nightingale and Flinn 2004). BlueFS is the first file system implemented for mobile computers with heterogeneous storage devices and an emphasis on energy efficiency. BlueFS is designed for a traditional file system workload. It replicates data across multiple storage devices on the mobile computing platform in the form of persistent file caches and supports disconnection on the storage device granularity in contrast to the platform granularity in file systems designed for mobility and weak connectivity. The primary copy resides at a file server.

The multiple device benefit is realized by directing file read operations to whatever is currently the lowest cost storage option. For example, if the hard drive is not spinning and network connectivity is strong, the low cost option may be remote storage; however, if the local disk is already spinning, the read may be directed to the disk. The system performs monitoring to maintain a running estimate of the current access time to each storage option and tracks the power state of each competing storage device. The ability to choose among devices from which to read implies a "read any, write to many" strategy to ensure that the desired data are available at each of the multiple storage options.

The "read any, write to many" scheme also implies that there will be many more write operations. The additional writes must be carefully managed so the energy savings by reads are not cancelled by the energy consumed to replicate the data across multiple storage devices. To accomplish this, writes are aggregated in a per-device buffer so that power state transitions can be amortized over a number of write operations for greater energy efficiency. Updates to the primary copy result in callbacks to invalidate cached copies on a per-device basis that are queued during disconnections, which are interpreted as states such as hibernation.

BlueFS addresses an important problem of coordinating multiple devices, each with its own dynamic power management based on observing request patterns for that individual device. The problem arises when a request is directed to the currently lowest cost device, although another of the alternatives might have been preferred if it had been in an active power state at the time. For example, if the disk is not spinning, a read request may be sent to remote storage via the wireless interface. Suppose this request is the first in a run of requests that could justify the disk spinup costs and make the disk the better choice. However, the disk power manager is normally not aware of the emerging access pattern because the requests are being diverted to the wireless interface. BlueFS uses ghost hints (Anand et al. 2004) to address this problem of unseen requests from the disk's point of view. Ghost hints capture the opportunity cost when a device was in the wrong power mode to be initially chosen, but might have been the ideal device otherwise. So, when the disk power manager receives enough ghost hints to justify the power state transition, it spins up the disk and, when the disk is ready, the request stream can be redirected to it. Figure 5.2 illustrates a scenario where the disk was not spinning but the network was active at the beginning of the request stream, initial requests are accessed over the network, a number of ghost hints trigger a disk spinup operation, and the network accesses hide the latency of the spinup operation. This latency hiding overlap capability is an additional advantage of multiple devices that can provide the same functionality with a request stream dynamically switched between them.

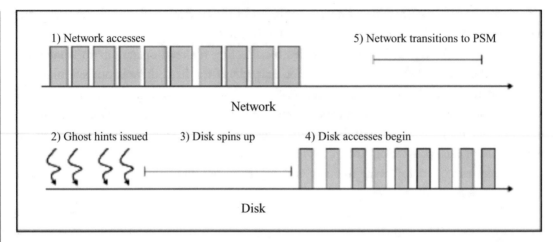

FIGURE 5.2: Initial network access hiding spinup latency and ghost hints triggering spinup (Nightingale and Flinn 2004).

5.2.3   Networking Alternatives

Choice among alternatives can also be applied to wireless network communication. Mobile devices such as PDAs and Smartphones are being equipped with multiple wireless interfaces that offer different capabilities. For example, devices supporting Bluetooth (for short range access), 802.11 WiFi (for local area communication), and GPRS (for wide area access) are currently on the market. The radios of these technologies differ significantly in power consumption, range, and bandwidth. Several projects have explored the possibility of exploiting the characteristics of multiple heterogeneous radios to reduce the energy consumption of wireless communication.

One of the widely recognized problems of wireless networking is that, while transmitting messages typically consume the most power, it is listening for incoming transmissions that may or may not arrive that actually consumes more energy (using lower power but for longer durations). Wake on Wireless (Shih et al. 2002) uses a lower power radio just as a wakeup channel to eliminate the need for the higher power radio to expend energy on listening. The lower power radio detects a signal indicating data are waiting to be delivered and wakes up the higher power radio to receive the data. The lower power radio does not need to support high bandwidth since it is not intended to actually carry data.

A mobile platform configured with both Bluetooth and WiFi wireless interfaces can transfer data on either radio and offers a choice between multiple devices able to serve in essentially similar roles. The idea is to allow dynamic switching back and forth between heterogeneous radios as conditions change to favor the characteristics of one over the other. Table 5.1 shows the different capabilities of specific examples of these radio technologies. Idle power shown in the table is the low-power, duty-cycled, listening mode for each

TABLE 5.1: Diversity of Radio Characteristics (Power as Reported in Datasheets)

RADIO TECHNOLOGY	RANGE	BANDWIDTH	TRANSMIT POWER	IDLE POWER
802.11 (Netgear MA701)	100 m	11 Mb/s	990 mW	264 mW
Bluetooth (BlueCore3)	10 m	1 Mb/s	81 mW	5.8 mW

technology—sniff mode for Bluetooth and PSM for WiFi. Notice that WiFi has the advantage of bandwidth and range, whereas Bluetooth has the advantage of low power.

The first requirement in realizing this concept is a mechanism that enables seamlessly switching a channel between interfaces. The second step is to discover effective policies for deciding when it is worthwhile for energy consumption to invoke a switch.

The mechanism needed for switching an IP channel between networking interfaces involves localized rerouting. One proposed scheme for localized network communication among neighbors is Contact Networking (Carter et al. 2003). Its goal is to provide all the support for connectivity among nearby nodes including neighbor discovery, localized naming, on-demand interface binding, channel management, and local routing. It uses link-layer awareness to provide lightweight discovery, rapid link failure detection, and optimizations that conserve resources. The architecture puts Contact Networking on the boundary between the link and network layers. Each node is uniquely identified by a global routable IP address (GRIP) that is simultaneously configured on all the node's interfaces, making the choice of interface into a routing problem.

Given the availability of multiple network interfaces and the mechanism to change a route from one to another, the next step is to come up with policies that decide when to switch up to a higher power/higher performance interface or down to a lower power/lower performance one in order to achieve energy savings. The CoolSpots project (Pering et al. 2006) is an investigation of some of the factors informing policies on a mobile device equipped with both WiFi and Bluetooth interfaces and a demonstration of the energy benefits of employing such policies. The goal is to reduce power consumption by appropriate choice of interface without compromising the needs of applications within a CoolSpot-enabled region as illustrated in Fig. 5.3.

Policies running on the mobile device must decide two things: when to switch up from the Bluetooth interface to WiFi (the Bluetooth radio remains on) and when to switch down to Bluetooth and turn off the WiFi radio. The process involves monitoring the network conditions that trigger the switch, activating or deactivating the WiFi interface, and notifying the base

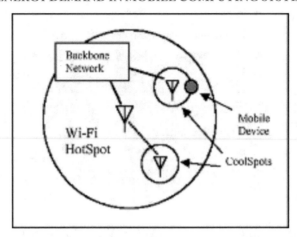

FIGURE 5.3: CoolSpot-enabled region. Bluetooth access points inside a WiFi hotspot (Pering et al. 2006) © 2006 ACM, Inc. Reprinted by permission. Courtesy Intel Research.

station. A switch incurs latency and energy overheads that may negate any benefit in the case of a poor policy decision. Several switching policies are considered. The bandwidth-based policy switches up or down based on the observed bandwidth crossing the same threshold in either direction. Since it is difficult to pick a static bandwidth threshold when channel capacity can change because of factors related to mobility, two other policies are proposed that explicitly measure capacity using a round trip time metric. The first of these capacity-based policies uses capacity to switch up to WiFi (e.g., RTT greater than 750 ms), but uses a static bandwidth threshold to switch down to Bluetooth. The second policy uses the same capacity-based criterion to switch up and a dynamic bandwidth threshold to switch down (the bandwidth at the time of the switch up is used as the threshold).

These policies are experimentally evaluated with a variety of benchmarks, various tuning parameters, and different locations away from the base station. The results confirm that the bandwidth-based policy has trouble with changing channel characteristics. Explicitly measuring capacity proves to adapt better. Averaging over all the benchmarks, the switching policies saved energy over the always-on WiFi with a small increase in latency. For streaming video applications, the dynamic capacity policy saves from 40% to 92% of the energy used by the fully active WiFi interface. Compared to using only Bluetooth across all the benchmarks, the switching policies performed significantly better in time with smaller energy savings (70–75% savings versus 85%).

5.2.4 Platform Tiers
The next step is to envision an integrated architecture that offers choices for computation and storage among several computing platforms with different power and performance character-

istics, packaged in one mobile device. The goal is to use the lowest power platform that can deliver the required functionality for the application. This is the idea being explored in the Turducken project (Sorber et al. 2005), focusing on applications that need to maintain local consistency with external data sources (e.g., an up-to-date view of email for the client residing on the mail server). The target is the energy used to periodically wake up a high power device just to check for updates.

The Turducken design is a hierarchy of platforms such that each tier is capable of waking up the tier above when the higher level tier is in a suspended state. All tiers share a common battery and communication links. Each tier contains its own independent processor, memory, and persistent storage system. There may be multiple radios or a single shared wireless interface. The user interacts with the device as a laptop (the superior tier), although any particular task can be distributed among tiers and executed by the most appropriate tier. High power tiers that are not needed to perform a task may go to sleep. Figure 5.4 shows the Turducken architecture.

A prototype of this architecture has been constructed and the energy savings measured with several data consistency applications. In the prototype, the tiers are still physically separate devices including a laptop, a PDA, and a sensor node. These heterogeneous platforms are shown in parentheses in each tier in Fig. 5.4. The lack of integration means that there are extra parts that could be eliminated (e.g., multiple displays). However, this prototype allows testing with the design concept. Three applications have been developed: time synchronization, web page caching, and email synchronization. Experiments have been done with three variants of the prototype: the base case of the laptop as a stand-alone device, a two-tier system combining the laptop and the sensor node, and the three-tier system that includes the PDA as a middle tier. Figure 5.5 summarizes some of the results with stacked bars showing the breakdown of power used by each tier in active and suspended states. Note the energy savings of the three-tier model. The two-tier version suffers from the overhead of the StrongARM suspend mode. These results show that the Turducken concept has the potential to dramatically extend the battery lifetime compared to a standard laptop for this class of applications.

The problem of distributing applications across tiers is an area that needs more attention for such architectures. In the experiments with the Turducken prototype, the applications have been tailored to the heterogeneous processors. There is currently no way to develop transparent software that can migrate tasks among tiers on a more dynamic basis.

5.3   PUTTING IT ALL TOGETHER—WHOLE SYSTEMS

The operating system resides in a unique position within a system where it can observe the resource demands of the mix of applications launched on the platform as well as the activity

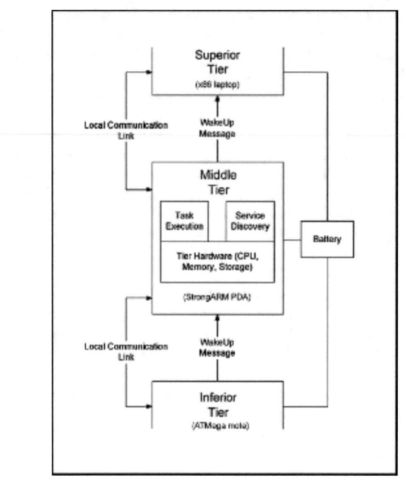

FIGURE 5.4: Three-tier Turducken architecture (with platforms used in prototype) (Sorber et al. 2005) © 2005 ACM, Inc. Reprinted by permission.

of the multiple hardware components as they consume power in their operation. This gives the OS a system-wide perspective on balancing the demand and supply of energy. Thus, the goal of the ECOSystem (Energy Centric Operating System) project (Zeng et al. 2003, 2005) is to explicitly manage energy as a first-class operating system resource and to understand the interactions with other resource management within a device. The project has been framed by two choices: to assume a general-purpose workload of applications that are not necessarily energy aware and to target the energy goal of achieving a specified battery lifetime. The challenge is to manage battery energy across time, distribute power among all the hardware devices that share the resource, and allocate energy fairly among multiple competing application demands. This creates a need for accurate accounting of energy use.

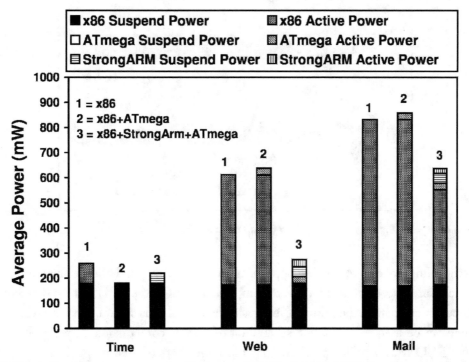

FIGURE 5.5: Turducken energy results (Sorber et al. 2005) © 2005 ACM, Inc. Reprinted by permission.

A new abstraction, "currentcy" (merging the words "current" and "currency"), is introduced to give energy a concrete representation within the system that allows it to be tracked, allocated, and scheduled like other OS resources. Applications can be granted a currentcy budget that can be spent to gain access to hardware resources that consume power on behalf of that application. The overall framework is illustrated in Fig. 5.6. In step 1, the overall allocation of currentcy per epoch of time is calculated from the target lifetime and remaining battery capacity. In step 2, the overall currentcy allocation is distributed among the competing tasks. Finally, in step 3, currentcy is deducted from a task's account as devices consume energy to meet the task's demands. Each step offers a variety of policy options to explore.

This currentcy framework has been implemented in the ECOSystem prototype, managing the CPU, hard disk, and wireless interface of a laptop via their use of energy. The prototype includes an embedded power model for each managed device, a task-tracking infrastructure to attribute device activity to the appropriate tasks, and device-specific payback policies to charge tasks for their energy consumption. Experiences with currentcy allocation policies in the prototype have considered ways to reclaim unused currentcy and have shown the need to

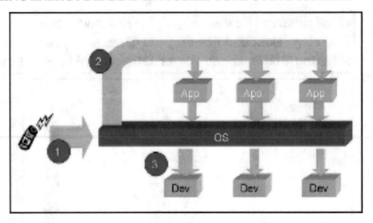

FIGURE 5.6: ECOSystem currentcy flow (Zeng et al. 2005).

align scheduling policies to provide opportunities to actually spend allocations once they are made. Scheduling tasks on one hardware component, such as the CPU, can be based on their overall rate of currentcy expenditure, even when the currentcy is spent on other devices. Thus, currentcy can be used to coordinate scheduling of various components. Experimental results with the ECOSystem prototype show that the target battery lifetime goal can be met and energy efficiency can be improved by employing system-wide management.

There are other whole-system energy management frameworks that have been developed including Nemesis (Neugebauer and McAuley 2001) and the Grace Project (Yuan and Nahrstedt 2003). Nemesis is also designed to meet a battery lifetime goal, charging processes for overuse when power consumption gets too high in order to encourage them to adapt their behavior. Grace is designed for a workload of soft real-time multimedia, taking an optimization approach to maximizing QoS, subject to battery lifetime constraints.

5.4   SUMMARY

In this chapter, we have expanded the focus from managing the energy consumption of a single device to consider interactions among multiple devices. We argue that these interactions may affect the underlying assumptions of the individual device policies.

Multiple devices may offer an opportunity to provide services in a more energy-efficient way depending on the current resource conditions. The idea is to provide different alternatives that are capable of supplying essentially the same functionality to the application at lower energy cost. One key to being able to effectively exploit the tradeoffs involving choices among multiple devices is good resource monitoring to determine what the network conditions and device power states are at the time.

Finally, we have discussed frameworks to manage the energy of an entire system that capture the dependencies among energy use of the various components on the platform. Accounting is again an important aspect in these frameworks. The whole system perspective elevates energy to a first-class resource to be managed by the operating system. In the next chapter, we broaden the view further to incorporate application-level knowledge into power management decision making.

CHAPTER 6

Energy-Aware Application Code

At several points in the previous chapters, we have alluded to projects that included a role for the applications to provide more information to assist the system in making power management decisions. Users may also want to respond with alternative behavior when energy supplies are low. In this chapter, we focus on techniques for application-level involvement. We first look at simple hints that the applications can provide about their usage patterns. Then we discuss actions that the applications can take to adapt their resource demand. Finally, we consider some challenges for application development and algorithm design specifically aimed at greater energy efficiency. The shaded portions of Fig. 6.1 show the new features covered in this chapter and provide context for the discussion.

6.1 APPLICATION INTERFACES TO ASSIST SYSTEM-LEVEL POWER MANAGEMENT

6.1.1 Usage Hints

In previous chapters, we have seen that one of the major challenges to power management has been the problem of accurately predicting the future resource demands of applications

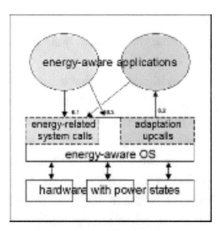

FIGURE 6.1: Application involvement in an energy-aware system.

running on the system. A recurring theme emerges that a little bit of information from applications about their usage patterns could potentially go a long way toward improving the effectiveness of system policies for both device power state transitions and voltage scaling. Such application-specific knowledge may still be only advisory since the operating system has the benefit of a more global view of the demands of the entire workload. There have been a number of examples of applications providing hints, via new system calls, about their current or future usage patterns to assist the system in making power management decisions.

For devices with discrete power states, the length of the upcoming idle gap between requests is vital information. There have been several projects addressing the timing of the next request for I/O on a hard disk drive. In Heath et al. (2002), a new system call is proposed to pass information about the next read request in an irregular access pattern. Irregular patterns are hard for the OS to easily predict from observations of prior idle periods. An example is an application that performs the same processing on every image in a directory. The knowledge about the size of the next image file could provide an estimate of the time until the next file access. The next_R system call, with an estimated idle time as an argument, allows the system to immediately spin down the disk on a sufficiently long idle gap and spin it back up before the next read will be issued. The value of the parameter may be found by profiling the first few iterations that generate disk reads to measure the average idle times. Similarly, Lu et al. (2002) offer a system call, RequireDevice (device, time, callback), for timer-based tasks that call a handler function (callback) on timer expiration. In this case, the time until processing by the callback function is determined by explicitly setting the timer. An example of its use is in an editor's autosave function that saves the working file every 5 min. The application can call RequireDevice (Harddrive, 5 min, savefile) to provide the OS with the information that this application will have a 5-min idle gap before needing the disk again. This may allow the OS to immediately spin down the disk and spin it back up just in time for the savefile function. These proposed system calls illustrate the benefits of relatively simple information flowing to the OS power manager.

The next kind of access pattern information that can be provided by an application to the system involves bracketing the beginning and end of an active period. As described in Section 3.3.2, the power management of an 802.11 wireless networking interface involves switching between power-saving mode (PSM) and continuously aware mode (CAM). The problem for system-level management is distinguishing whether the observable data transmission rate is an artifact of the beacon period or the natural behavior of the application. Hints from the application can convey the user's intent in accessing the wireless network (Anand et al. 2003, Kravets and Krishnan 2000). In self-tuning power management (Anand et al. 2003), the application interface includes TransferHintBegin with parameters indicating whether a forthcoming transfer is related to foreground or background activity. This is useful since interactive applications are particularly sensitive to latency. The hint may also give an expected amount of data. This allows the system to transition to CAM if the data size exceeds the breakeven size or it is interactive. A matching HintEnd informs the system that this transmission is over and, from this application's point of view, it is OK to transition back into PSM. Similarly, ListenHintBegin – HintEnd calls can bracket a listening session and specify the maximum delay that can be tolerated. Finally, SetKnob specifies the relative importance to the application of performance versus energy conservation. A similar kind of hint is suggested in Flautner et al. (2001) for the problem of automatically detecting episodes for dynamic voltage scaling (Section 4.5). The proposed API consists of episode_begin and episode_end system calls to bracket an instance of an interactive, periodic, or producer–consumer episode and to have the option of specifying a deadline.

File caching and prefetching are discussed in Section 3.3.1 as a technique for shaping the disk request patterns, lengthening idle periods to provide more opportunities for spinning down the disk. One of the challenges of prefetching data is deciding which data and how much to prefetch in one active burst. In Papathanasiou and Scott (2004), these decisions are guided by hints provided by the application through a system call interface. The hint interface allows the application to explicitly specify the overall access pattern (sequential, loop, or random) and the estimated time of first and last access for a file. If the access pattern is deemed to be random, the interface allows the application to give a list of "hot" data clusters, the probability of their access, and followed-by probabilities to derive likely sequences. As an alternative, a monitoring daemon can generate hints on behalf of the application. These automatically generated hints are based on a database of file access patterns that have been monitored and saved from previous runs of the application. These saved profiles of application-specific behavior are targeted at those applications whose access patterns have a significant impact on disk energy efficiency. The system is depicted in Fig. 6.2.

Hints have also been used in dynamic voltage scheduling (DVS), as described in Section 4.4.1 where we described the use of these hints by the OS in setting the speed and voltage during periodic interrupts called power management points (PMPs) (AbouGhazaleh et al. 2006). The off-line compiler analysis annotates the program by inserting code that updates a designated memory location, the WCR. The value written into WCR represents the worst-case remaining cycles in the task. It conveys to the operating system the progress of the application and any dynamic slack that is accruing based on runtime control flow. The hint-generating code, called a power management hint or PMH, computes a new value for WCR based on its previous value and the actual path taken to arrive at this PMH. The placement of PMHs must ensure that the WCR value is updated at least once during every interval between PMPs.

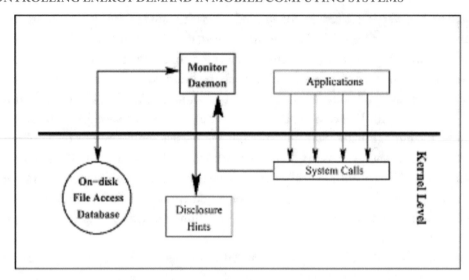

FIGURE 6.2:   Combining hints from system calls issued by the application and profiles of past access patterns to generate hints for prefetching (Papathanasiou and Scott 2004).

The placement of PMHs is performed in three steps. The first step is to form regions in the control flow graph that aggregate one or more basic blocks while maintaining the structure of the program. Figure 6.3 shows a control flow graph broken into five regions. Next, profiling is performed to capture timing information about each region, for each procedure, and the cycle count and maximum number of iterations for each loop. Then the analysis determines the PMP interval length (in cycles) that will be communicated to the system along with the location of the WCR via a system call (inserted by the compiler) during process initialization. Finally, the placement algorithm traverses the control flow graph, maintaining a cycle counter, ac, and inserting a PMH before the ac exceeds the PMP interval.

6.1.2   System Calls for More Flexibility in Timing

Beyond hints that convey usage patterns and process behavior to the system, there are usage scenarios in which the application may be able to grant the system more flexibility in servicing requests. APIs that specify an acceptable tolerance on performance degradation can be exploited by the system in its scheduling of idle and busy periods. For example, the API proposed in Anand et al. (2003) includes a maximum delay parameter in the ListenHintBegin call.

A variant of the RequireDevice system call in Lu et al. (2002) specifies a tolerance range around the timer value. This is used by the system in process scheduling decisions for tasks invoked through the callback mechanism on timer expiration. Flexible timers allow the execution order of tasks to be rearranged, within their specified tolerances, to make idle periods on the device longer. The scheduler can also group a task's execution with other tasks requiring

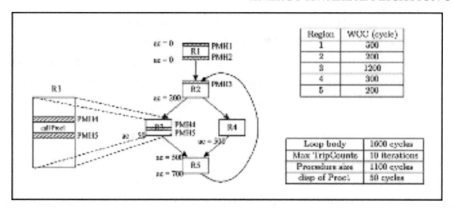

FIGURE 6.3: Compiler insertion of program management hints (PMH) into the control flow graph. The profiled timings in cycles affect the setting of the interrupt interval, the placement of PMHs, and the computation of the WCR hint value at each PMH (AbouGhazaleh et al. 2006) © 2006 ACM, Inc. Reprinted by permission.

the same device to be active within the same timeframe to amortize power state transitions among the group. This may enable the power management system to employ deeper sleep modes on the devices.

Another interesting example is the set of cooperative I/O system calls proposed in Weissel et al. (2002). An application can use CoopIO to essentially give the OS its permission to reshape disk request patterns. The cooperative versions of the basic file open, read, and write operations include a timeout and a cancel flag as additional parameters. The specified timeout allows the request to wait if the disk is not already spinning until another I/O request arrives or the timeout expires to trigger a spinup operation. If the cancel flag is set when the timeout expires and no other I/O request has arrived to cause the disk to spin up, then the CoopIO request can be canceled altogether. The benefits to the system are that idle disks can remain in the low power state longer. Applications may choose to use CoopIO features for logging data, periodic requests on multimedia files, and autosave functions. The possible delay on I/O operations suggests that a separate thread may be useful for issuing cooperative operations. An autosave thread is a good example to illustrate the use of a write request that can be cancelled. A pending write is unnecessary when another autosave period is already due to expire and will just issue the next write.

6.2   OS-APPLICATION INFORMATION FLOW TO ENABLE ADAPTATION

6.2.1   Frameworks for System Feedback

So far, we have considered information flowing from the application to the system about usage patterns to inform power management policies. In this section, we introduce information flow from the system to the application that may cause the application to adapt its behavior in response and the system mechanisms providing such two-way exchanges.

Microsoft has introduced the OnNow initiative (Microsoft Corp 2001) to establish the operating system's central role in power management and to encourage collaboration between applications and the operating system toward improving their power use. The OnNow architecture is built upon the Advanced Configuration and Power Interface (ACPI) specification that standardizes the system model based on global and device power states, as discussed in Section 2.1.2. With OnNow, the OS assumes responsibility for coordinating power management at all levels. In particular, it defines a new application interface (incorporated into the Win32 API and extended in Vista) that exposes power management capabilities to software developers and enables them to participate in designing more energy-aware applications. The API features include mechanisms for applications to inform the OS of application requirements and activity that may not be detected by the OS. For example, the SetThreadExecutionState call can be used to indicate that the display is needed regardless of the lack of user interface events that usually signify an idle machine. This can be used to solve the annoying problem of the display going blank during a slide show presentation.

Information flow in the other direction, from the system to the applications, is equally valuable. WM_POWERBROADCAST messages are used to notify all running applications of power management events such as the system being put into a sleep state. Applications have the opportunity to properly respond to such events by cleanly closing down and preserving the user's work appropriately. There are also power status functions that allow applications to perform differently based on the power source or on device power states. The GetSystemPowerStatus call returns information on whether the power source is AC or DC and how much battery lifetime is remaining. An energy-aware application may use this information to postpone maintenance tasks when the machine is running on batteries or restrict certain activities when the battery power is running low. The GetDevicePowerState call returns true or false depending on whether the specified device is fully operational. An example using this feature is an application that defers lower priority disk I/O while the disk is not already spinning. The goal of OnNow is to provide the support to foster creative energy-aware application development for the Windows environment.

Another framework offering a feedback mechanism to applications about their energy consumption is the Nemesis OS (Neugebauer and McAuley 2001). Nemesis emphasizes accurate energy accounting for both processes and devices. The proposed model for the collaborative relationship between applications and the OS is based on economic ideas. Accurate accounting provides the basis to charge processes a tax for excessive energy consumption. This is defined in terms of the limit on the battery discharge rate required to reach a desired battery lifetime. Charging processes in proportion to their excess energy consumption provides a useful feedback signal. Since individual processes are allocated a limited number of energy credits to spend on energy use, these charges provide an economic incentive for the application to adapt its behavior to use less energy. Techniques for actually performing adaptation in response to these charges are not addressed in this work.

6.2.2   Adaptation Through Fidelity

The Odyssey system (Flinn and Satyanarayanan 1999b, 2004) addresses both a system framework and techniques that applications can use to dynamically adapt their behavior. The main abstraction is data *fidelity*. Application adaptation involves reducing fidelity of a data object or degrading data quality relative to the original object in application-specific ways to reduce the energy consumption of transmitting and/or processing the data. Examples of lower fidelity data include digital maps with certain features (e.g., rivers, businesses, minor roads) suppressed, images cropped to a smaller size, a video displayed in black and white instead of the original full color version, and images distilled using lossy compression. What constitute acceptable fidelity-reducing transformations depend both on the type of data and on the needs of the application using them.

The Odyssey architecture supports application-specific adaptation. Figure 6.4 shows the components of the architecture on the mobile client. The viceroy is responsible for monitoring resource availability, in particular, the estimate of remaining battery lifetime and rate of consumption. The wardens provide type-specific operations for producing data at different fidelity levels. The API allows applications to specify resource needs and to register the fidelity levels and type-specific operations to be supported. Odyssey provides notifications via an upcall to

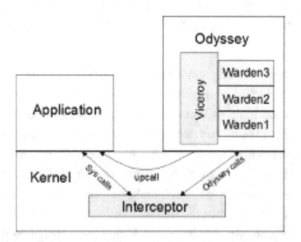

FIGURE 6.4: The Odyssey architecture (Flinn and Satyanarayanan 1999b, 2004) © 1999 ACM, Inc. Reprinted by permission.

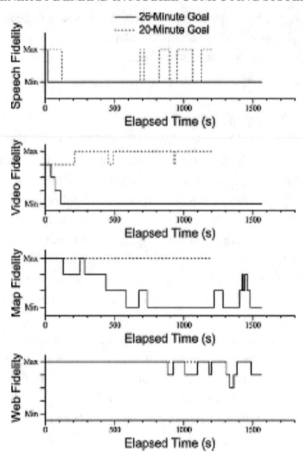

FIGURE 6.5: Change in fidelity levels of four applications for battery lifetimes of 20 and 26 minutes (Flinn and Satyanarayanan 1999b, 2004) © 2004 ACM, Inc. Reprinted by permission.

the applications when expectations are not being met. Upon receiving notifications, an application can adjust its data fidelity to match the new expectations on resource availability. Thus, if the power demand is too high, Odyssey triggers adaptations by issuing notifications to the applications. Conversely, if the battery capacity exceeds demand, applications may be notified so that they can increase fidelity to improve the user's experience with higher quality data.

Figure 6.5 illustrates Odyssey causing adaptations in applications in order to reach a particular battery lifetime goal. These are the results of an experiment with concurrent applications: A browser accesses images from the web with different JPEG quality factors to yield five supported fidelity levels. The map data have four fidelity levels including the full map, filtering of minor roads, filtering of secondary roads, and a combination of filtering and cropping. The video application supports four levels of fidelity including the full quality video, two levels of lossy compression, and a combination of compression and a reduced viewing window. The speech recognition supports two levels of fidelity related to the size of the vocabulary used. In these experiments, Odyssey is using an incremental, priority-based policy of notifications to trigger adaptations. To make the experiments reasonably short, the energy supply is set to last only 19:27 min at the highest fidelities. The graphs show the different applications adjusting their fidelity levels in response to the battery lifetime goals of either 20 or 26 min.

6.3 DEVELOPING APPLICATIONS FOR ENERGY EFFICIENCY

There has been considerable research devoted to either compiler transformations of ordinary programs to yield more energy-efficient code or algorithm design efforts to directly develop low-power code that does not rely on special APIs or the support of an energy-aware OS. These efforts usually target an abstract energy model of the hardware platform. Sometimes manual transformations are viewed as a first step toward eventually automating energy-saving techniques. For example, one compiler project aims to cluster array layouts within physical memory to optimize placement for power-aware memory chips and generate code to perform power state transitions of unused memory modules under program control (Delaluz et al. 2001). This work assumes a single program environment and no virtual memory. Similarly, compiling approaches have been proposed to embed speed and voltage scaling points directly into program code (e.g., Shin and Kim 2001). The potential pitfall of such techniques lies in ignoring and, therefore, conflicting with the management efforts of the OS. Explicitly designing application programs to be low power requires the developer to have a good energy model of the platform and an understanding of the resource management being done in other system layers. This is why APIs like OnNow that encourage OS/application cooperation are attractive. The OS has a system-wide runtime perspective that purely program-directed strategies would lack.

6.4 SUMMARY

In this chapter, we have made a case that even a little bit of semantic information about the user-level application is valuable to the system. The application's intentions in using the resource may not match the system's default management assumptions. Information exchange across layers, at various levels and in both upward and downward directions, can have a positive role to play in discovering new approaches to manage energy demand.

CHAPTER 7

Challenges and Opportunities

There is widespread recognition within the mobile computing R&D community of the importance of energy management. This lecture has focused on the demand side of the energy story. It has highlighted some of the significant contributions that have been made to our understanding of how to manage device power consumption and influence workload demand to save energy and prolong battery life. There is a growing appreciation that further advances in energy/power management are needed bridging all levels of system design—from the hardware to the applications. In this chapter, we recap some of the lessons learned from experience so far and consider the challenges ahead.

7.1   ON IMPROVING HARDWARE CAPABILITIES

The availability of hardware features, such as low power modes and voltage/frequency scaling, invites system efforts to explore how to creatively exploit these capabilities. These efforts have, in turn, provided feedback on desirable improvements in the hardware that could make the solutions that employ those features more effective.

Some of the major constraints on power management algorithms have been the properties of the low power states provided and the transition costs to enter those states and return from them. From the hardware side, the characteristics of the available power states are determined by the circuitry that can be selectively and incrementally disabled and by the operations required to restore full activity. From the systems perspective, the transition overheads, in particular, may severely limit the cases in which low power modes can be used. The breakeven times of the available low power states may be poorly matched to the idle gaps in the device access patterns. Consequently, certain power states may not represent viable choices for policies to even consider. Efforts by hardware developers to reduce transition costs can enable more aggressive policies that employ power states. Similarly, experiments with discrete combinations of voltage and frequency supported by some early scalable processors have identified unproductive choices. Ideally, the hardware needs to offer software policies a sufficiently broad range of useful settings to exploit.

Another limitation on achieving system-wide energy savings has been the relatively high level of base power consumption that is needed on some platforms regardless of power management (e.g., to be responsive to wakeup events). Reducing the base power consumption is another desirable goal for hardware improvement.

Finally, in order to enable the sophisticated policies that are possible within the operating system, the hardware capabilities must be exposed to software control. Software policy making can also be enhanced with more information about the power consumption of individual components. The smart battery interface is a move in that direction, but finer-grain instrumentation built into the hardware (e.g., the power analogue to processor performance counters) may be attractive from the systems perspective.

7.2   ON SYSTEM SOFTWARE AS A FOCAL POINT

There is a compelling argument that the operating system is the appropriate layer to serve as the center of power management, although all layers ultimately must be involved. The OS has a global and dynamic view of the access patterns generated by the workload running on the machine as well as the hardware utilization. On the other hand, there are limits to the ability of the OS to predict resource demands. We have seen that the system can benefit significantly from a bit of extra knowledge about the applications. Thus, information flow across boundaries needs to be encouraged and appropriate interfaces need to be defined.

Power management for individual devices has been extensively studied with areas such as dynamic voltage scheduling for real-time processes and spindown policies for disks becoming relatively mature. Many of the studies have evaluated proposed solutions through simulation. However, experiences with prototype implementations have not always achieved the predicted levels of energy savings because of interactions with the rest of the system. Taking a whole-system approach is increasingly important in moving forward.

7.3   ON ENGAGING APPLICATION DESIGNERS AND USERS

In general, application programs have been developed with little or no attention paid to their energy efficiency. It is easy to find examples of wasteful activity (e.g., blinking cursors, screen-savers) or poor timing (e.g., auto-updates that fill idle gaps). There is tremendous opportunity to improve energy consumption by rewriting programs to eliminate waste and configuring them so as to not work against the system's power management. Some of this can be achieved by simply promoting more awareness among application developers and users of power management capabilities that already exist and how to employ them. The human–computer interaction (HCI) community might be engaged in designing more energy-efficient user interfaces. One idea may be to modify the feedback methods on program progress that are commonly used.

Creating truly energy-aware applications represents a new direction for most programmers. By working cooperatively with the OS, application design can have a significant impact on energy use. Energy-aware applications on mobile devices can deliver improved battery life which should be a competitive advantage. As in any application-level optimization, the developer needs to work from a good understanding of the system-level power management model in order to complement rather than compete with the system's efforts.

7.4   GOING FORWARD

Several trends point to an increasing need for expertise in energy management and energy-aware software development. One trend is the growing reliance on mobile, wireless devices in our professional and personal lives. Another is the rapid evolution of new usage scenarios for these devices. These will place increasing demands on the battery resources. Another trend is society's growing awareness of the role of energy in climate change. Energy conservation will become a higher priority as we look for ways to reduce the size of our carbon footprint. The skills required in developing more energy-efficient computing products and services will be valued not only in the mobile computing field, but extending into the computing discipline as a whole.

References

AbouGhazaleh, N., Mossé, D., Childers, B. R., and Melhem, R., "Collaborative operating system and compiler power management for real-time applications," *Trans. Embedded Comput. Syst.*, Vol. 5, No. 1, pp. 82–115, Feb. 2006. doi:10.1145/1132357.1132361

Anand, M., Nightingale, E. B., and Flinn, J., "Self-tuning wireless network power management," in *Proc. 9th Annual Conf. Mobile Computing and Networking (MOBICOM '03)*, San Diego, CA, September, 2003, pp. 176–189.

Anand, M., Nightingale, E. B., and Flinn, J., "Ghosts in the machine: interfaces for better power management," in *Proc. 2nd Int. Conf. Mobile Systems, Applications, and Services (MobiSys '04)*, Boston, MA, USA, June 06–09, 2004, pp. 23–35.

Aydin, H., Melhem, R., Mossé, D., and Mejía-Alvarez, P., "Power-aware scheduling for periodic real-time tasks," *IEEE Trans. Comput.*, Vol. 53, No. 5, pp. 584–600, May 2004. doi:10.1109/TC.2004.1275298

Bellosa, F., "The benefits of event: driven energy accounting in power-sensitive systems," in *Proc. 9th Workshop on ACM SIGOPS European Workshop: Beyond the Pc: New Challenges for the Operating System (EW 9)*, Kolding, Denmark, September 17–20, 2000, pp. 37–42.

Brooks, D., Tiwari, V., and Martonosi, M., "Wattch: a framework for architectural-level power analysis and optimizations," in *27th Annual Int. Symp. on Computer Architecture (ISCA)*, June 2000, pp. 83–94.

Carter, C., Kravets, R., and Tourrilhes, J., "Contact networking: a localized mobility system," In *Proc. 1st Int. Conf. Mobile Systems, Applications and Services (MobiSys '03)*, San Francisco, CA, May 05–08, 2003, pp. 145–158.

Chandra, S. and Vahdat, A., "Application-specific network management for energy-aware streaming of popular multimedia formats," in *Proc. General Track: 2002 USENIX Annual Technical Conf.*, June 10–15, 2002, pp. 329–342.

Chen, J. W., Dubois, M., and Stenstrm, P., "Integrating complete-system and user-level performance/power simulators: the SimWattch approach," in *Proc. Int. Symp. on Performance Analysis of Systems and Software*, 2003, pp. 1–10.

Chiasserini, C. F. and Rao, R. R., "Pulsed battery discharge in communication devices," in *Proc. 5th Annual ACM/IEEE Int. Conf. on Mobile Computing and Networking (MobiCom '99)*, Seattle, Washington, DC, August 15–19, 1999, pp. 88–95.

Childers, B., Tang, H., Melhem, R., "Adapting processor supply voltage to instruction-level parallelism," *Koolchips 2000 Workshop during MICRO-33*, 2000.

Contreras, G., Martonosi, M., Peng, J., Ju, R., and Lueh, G., "XTREM: a power simulator for the Intel XScale® core," in *Proc. 2004 ACM SIGPLAN/SIGBED Conf. Languages, Compilers, and Tools For Embedded Systems (LCTES '04)*, Washington, DC, June 11–13, 2004, pp. 115–125.

Contreras, G. and Martonosi, M., "Power prediction for Intel XScale® processors using performance monitoring unit events," in *Proc. 2005 Int. Symp. on Low Power Electronics and Design (ISLPED '05)*, San Diego, CA August 08–10, 2005, pp. 221–226.

Dalton, A. and Ellis, C., "Sensing user intention and context for energy management," in *Proc. Hot Topics in Operating Systems (HOTOS)*, May 2003.

Delaluz, V., Kandemir, M., Vijaykrishnan, N., Sivasubramaniam, A., and Irwin, M. J., "Hardware and software techniques for controlling DRAM power modes," *IEEE Trans. Comput.*, Vol. 50, No. 11, pp. 1154–1173, Nov. 2001. doi:10.1109/12.966492

Delaluz, V., Sivasubramaniam, A., Kandemir, M., Vijaykrishnan, N., and Irwin, M. J. "Scheduler-based DRAM energy management," in *Proc. 39th Conf. on Design Automation (DAC '02)*, New Orleans, LA, June 10–14, 2002, pp. 697–702.

Douglis, F., Krishnan, P., and Bershad, B. N., "Adaptive disk spin-down policies for mobile computers," in *Proc. 2nd Symp. on Mobile and Location-Independent Computing*, April 10–11, 1995, pp. 121–137.

Fan, X., Ellis, C., and Lebeck, A., "The synergy between power-aware memory systems and processor voltage," in *Proc. Power-Aware Computer Systems (PACS)*, 2003.

Flautner, K., Reinhardt, S., and Mudge, T., "Automatic performance setting for dynamic voltage scaling," in *Proc. 7th Annual Int. Conf. Mobile Computing and Networking (MobiCom '01)*, Rome, 2001, pp. 260–271.

Flautner, K. and Mudge, T., "Vertigo: automatic performance-setting for Linux," in *Proc. 5th Symp. on Operating Systems Design and Implementation (OSDI '02)*, Boston, MA, December 09–11, 2002, pp. 105–116.

Flinn, J. and Satyanarayanan, M., "PowerScope: a tool for profiling the energy usage of mobile applications," in *Proc. 2nd IEEE Workshop on Mobile Computer Systems and Applications (wmcsa)*, 1999a, pp. 2–10.

Flinn, J. and Satyanarayanan, M., "Energy-aware adaptation for mobile applications," in *Proc. 17th ACM Symp. on Operating Systems Principles (SOSP '99)*, Charleston, SC, December 12–15, New York: ACM, 1999b, pp. 48–63.

Flinn, J., Park, S., and Satyanarayanan, M., "Balancing performance, energy, and quality in pervasive computing," in *Proc. 22nd Int. Conf. Distributed Computing Systems (ICDCS'02)*, July 02–05, Washington, DC: IEEE Computer Society, 2002, pp. 217–226.

Flinn, J. and Satyanarayanan, M., "Managing battery lifetime with energy-aware adaptation," *ACM Trans. Comput. Syst.*, Vol. 22, No. 2, pp. 137–179, May 2004. doi:10.1145/986533.986534

Grunwald, D., Levis, P., Neufeld, C., and Farkas, K., "Policies for dynamic clock scheduling," in *Proc. 4th USENIX Symp. on Operating Systems Design and Implementation (OSDI 2000)*, San Diego, CA, October 2000, pp. 73–86.

Gurumurthi, S., Sivasubramaniam, A., Irwin, M. J., Vijaykrishnan, N., Kandemir, M., Li, T. and John L., "Using complete machine simulation for software power estimation: the softwatt approach," in *Proc. 8th Int. Symp. on High Performance Computer Architecture*, 2002, pp. 141–150.

Gurumurthi, S., Sivasubramaniam, A., Kandemir, M., and Franke, H., "DRPM: dynamic speed control for power management in server class disks," in *Proc. 30th Int. Symp. on Computer Architecture (ISCA '03)*, San Diego, CA, June 09–11, New York: ACM, 2003, pp. 169–181.

Gurun, S. and Krintz, C., "AutoDVS: an automatic, general-purpose, dynamic clock scheduling system for hand-held devices," in *Proc. 5th ACM Int. Conf. on Embedded Software (EMSOFT '05)*, Jersey City, NJ, September 18–22, 2005, New York: ACM, 2005, pp. 218–226.

Heath, T., Pinheiro, E., Hom, J., Kremer, U., and Bianchini, R., "Application transformations for energy and performance-aware device management," in *Proc. 2002 Int. Conf. Parallel Architectures and Compilation Techniques (PACT)*, September 22–25, Washington, DC: IEEE Computer Society, 2002, pp. 121–130.

Helmbold, D. P., Long, D. D., and Sherrod, B. "A dynamic disk spin-down technique for mobile computing," in *Proc. 2nd Annual Int. Conf. Mobile Computing and Networking (MobiCom '96)*, Rye, New York, New York: ACM, 1996, pp. 130–142.

Huang, H. and Pillai, P., and Shin, K. G., "Design and implementation of power-aware virtual memory," in *USENIX Annual Technical Conf.*, 2003, pp. 57–70.

Hwang, C. and Wu, A. C., "A predictive system shutdown method for energy saving of event-driven computation," in *Proc. 1997 IEEE/ACM Int. Conf. Computer-Aided Design*, San Jose, CA, November 09–13, Washington, DC: IEEE Computer Society, 1997, pp. 28–32.

Intel Corporation, Advanced Configuration and Power Interface (ACPI), 2000.

Irani, S., Shukla, S., and Gupta, R., "Online strategies for dynamic power management in systems with multiple power-saving states," *ACM Trans. Embedded Comput. Syst.*, Vol. 2, No. 3, pp. 325–346, Aug. 2003. doi:10.1145/860176.860180

Iyer, S., Luo, L., Mayo, R., and Ranganathan, P., "Energy-adaptive display system designs for future mobile environments," in *Proc. 1st Int. Conf. Mobile Systems, Applications and Services*, 2003, pp. 245–258.

Joseph, R. and Martonosi, M., "Run-time power estimation in high performance microprocessors," in *Proc. 2001 Int. Symp. on Low Power Electronics and Design (ISLPED '01)*, Huntington Beach, CA, 2001, pp. 135–140.

Kistler, J. J. and Satyanarayanan, M., "Disconnected operation in the Coda file system," in *Proc. 13th ACM Symp. on Operating Systems Principles (SOSP '91)*, Pacific Grove, CA, October 13–16, 1991, pp. 213–225.

Kravets, R. and Krishnan, P., "Application-driven power management for mobile communication," *Wirel. Netw.*, Vol. 6, No. 4, pp. 263–277, Jul. 2000. doi:10.1023/A:1019149900672

Kuenning, G. and Popek, G., *"Automated hoarding for mobile computers,"* in *Proc. 16th ACM Symp. on Operating Systems Principles (SOSP-16)*, St. Malo, France, October 5–8, 1997, pp. 264–275.

Kotz, D. and Essien, K., "Analysis of a campus-wide wireless network," *Wirel. Netw.*, Vol. 11, No. 1–2, pp. 115–133, Jan. 2005.

Krashinsky, R. and Balakrishnan, H., "Minimizing energy for wireless web access with bounded slowdown," in *Proc. 8th Annual Int. Conf. Mobile Computing and Networking (MobiCom '02)*, Atlanta, GA, September 23–28, 2002, pp. 119–130.

Krishnan, P., Long, P., and Vitter, J. "Adaptive disk spin-down via optimal rent-to-buy in probabilistic environments," *Algorithmica*, Vol. 23, No. 1, pp. 31–56, Jan. 1999. doi:10.1007/PL00009249

Lebeck, A. R., Fan, X., Zeng, H., and Ellis, C., "Power aware page allocation," in *Proc. 9th Int. Conf. Architectural Support for Programming Languages and Operating Systems (ASPLOS-IX)*, Cambridge, MA, 2000, pp. 105–116.

Li, K., Kumpf, R., Horton, P., and Anderson, T., "A quantitative analysis of disk drive power management in portable computers," in *Proc. USENIX Winter Technical Conf.*, 1994, pp. 279–291.

Lorch, J. R. and Smith A. J., "Improving dynamic voltage scaling algorithms with PACE," in *Proc. 2001 ACM SIGMETRICS Int. Conf. Measurement and Modeling of Computer Systems (SIGMETRICS '01)*, Cambridge, MA, 2001, pp. 50–61.

Lu, Y., Benini, L., and De Micheli, G., "Power-aware operating systems for interactive systems," *IEEE Trans. Very Large Scale Integr. Syst.*, Vol. 10, No. 2, pp. 119–134, Apr. 2002. doi:10.1109/92.994989

Magnusson, P., Christensson, M., Eskilson, J., Forsgren, D., Hallberg, G., Hogberg, J., Larsson, F., Moestedt, A., and Werner, B., "Simics: a full system simulation platform," *Computer*, Vol. 35, No. 2, pp. 50–58, Feb. 2002. doi:10.1109/2.982916

Mahesri, A. and Vardhan, V., "Power consumption breakdown on modern laptop," in *Proc. Workshop on Power-Aware Computing Systems*, Portland OR, December 2004.

Martin, T. L. and Siewiorek, D. P., "Nonideal battery and main memory effects on CPU speed-setting for low power," *IEEE Trans. Very Large Scale Integr. Syst.*, Vol. 9, No. 1, pp. 29–34, Feb. 2001. doi:10.1109/92.920816

Microsoft Corp., "OnNow Power Management Architecture for Applications," https://www.microsoft.com/whdc/archive/OnNowApp.mspx, 2001.

Neugebauer, R. and McAuley, D., "Energy is just another resource: energy accounting and energy pricing in the nemesis OS," in *Proc. 8th Workshop on Hot Topics in Operating Systems (HOTOS)*, May 20–22, 2001, p. 67.

Nightingale, E. B., and Flinn, J., "Energy-efficiency and storage flexibility in the Blue File System," in *Proc. 6th Symposium on Operating Systems Design and Implementation*, San Francisco, CA, December 2004, pp. 363–378.

Papathanasiou, A. and Scott, M., "Energy efficient prefetching and caching," in *Proc. USENIX 2004 Annual Technical Conf. (USENIX '04)*, June 27–July 2, Boston, MA: Boston Marriott Copley Place, 2004, pp. 255–268.

Peek, D. and Flinn, J., "Drive-thru: fast, accurate evaluation of storage power management," in *Proc. USENIX Annual Technical Conf.*, Anaheim, CA, April 2005, pp. 251–264.

Pering, T., Agarwal, Y., Gupta, R., and Want, R., "*CoolSpots*: reducing the power consumption of wireless mobile devices with multiple radio interfaces," in *Proc. 4th Int. Conf. Mobile Systems, Applications and Services (MobiSys 2006)*, Uppsala, Sweden, June 19–22, 2006, pp. 220–232.

Pillai, P. and Shin, K. G., "Real-time dynamic voltage scaling for low-power embedded operating systems," in *Proc. 18th ACM Symp. on Operating Systems Principles (SOSP '01)*, Banff, AB, Canada, October 21–24, 2001, pp. 89–102.

Rosenblum, M., Bugnion, E., Devine, S., and Herrod, S. A., "Using the SimOS machine simulator to study complex computer systems," *ACM Trans. Model. Comput. Simul.*, Vol. 7, No. 1, pp. 78–103, Jan. 1997. doi:10.1145/244804.244807

Rudenko, A., Reiher, P., Popek, G. J., and Kuenning, G. H., "The remote processing framework for portable computer power saving," in *Proc. 1999 ACM Symp. on Applied Computing (SAC '99)*, San Antonio, TX, February 28–March 02, New York: ACM, 1999, pp. 365–372.

Shnayder, V., Hempstead, M., Chen, B., Allen, G. W., and Welsh, M., "Simulating the power consumption of large-scale sensor network applications," in *Proc. 2nd Int. Conf. Embedded Networked Sensor Systems (SenSys '04)*, Baltimore, MD, November 03–05, 2004, pp. 188–200.

Shih, E., Bahl, P., and Sinclair, M. J., "Wake on wireless: an event driven energy saving strategy for battery operated devices," in *Proc. 8th Annual Int. Conf. Mobile Computing and Networking (MobiCom '02)*, Atlanta, GA, September 23–28, 2002, pp. 160–171.

Shin, D. and Kim, J., "A profile-based energy-efficient intra-task voltage scheduling algorithm for real-time applications," in *Proc. 2001 Int. Symp. on Low Power Electronics and Design (ISLPED '01)*, Huntington Beach, CA, 2001, pp. 271–274.

Šimunic', T., Benini, L., and De Micheli, G., "Cycle-accurate simulation of energy consumption in embedded systems," in *Proc. 36th ACM/IEEE Conf. Design Automation*, June 1999, pp. 867–872.

Simunic, T., Benini, L., Glynn, P., and De Micheli, G., "Dynamic power management for portable systems," in *Proceedings of the 6th Annual Int. Conf. Mobile Computing and Networking (MobiCom '00)*, Boston, MA, August 06–11, 2000, pp. 11–19.

Singh, S. and Raghavendra, C., "Power efficient MAC protocol for multihop radio networks," in *Proc. IEEE PIRMC'98*, vol. 1, September, 1998, pp. 153–157.

Sorber, J., Banerjee, N., Corner, M. D., and Rollins, S., "Turducken: hierarchical power management for mobile devices," in *Proc. 3rd Int. Conf. Mobile Systems, Applications, and Services (MobiSys '05)*, Seattle, Washington, June 06–08, 2005, pp. 261–274.

Tan, T., Raghunathan, A., and Jha, N., "EMSIM: an energy simulation framework for an embedded operating system," in *Proc. Int. Symp. Circuit & Systems*, May, 2002, pp. 464–467.

Tiwari, V., Malik, S., Wolfe, A., and Lee, M., "Instruction level power analysis and optimization of software," *J. VLSI Signal Process.*, pp. 1–18, Vol. 13, No. 2–3, 1996.

Ye, W., Vijaykrishnan, N., Kandemir, M., and Irwin, M. J., "The design and use of simplepower: a cycle-accurate energy estimation tool," in *Proc. 37th Conf. Design Automation (DAC '00)*, Los Angeles, CA, June 05–09, Vol. 3, No. 2–3, 2000, pp. 340–345.

Weiser, M., Welch, B., Demers, A., and Shenker, S., Scheduling for reduced CPU energy, Proc. 1st Symp. on Operating Systems Design and Implementation, Nov. 1994, pp. 13–23.

Weissel, A., Beutel, B., and Bellosa, F., "Cooperative I/O: a novel I/O semantics for energy-aware applications," Originally Published in the Proceedings of OSDI '02: 2nd Symp. on Operating Systems Design and Implementation (Berkeley, CA: USENIX Association, 2002), pp. 117–129.

Xu, R., Xi, C., Melhem, R., and Mosse, D., "Practical PACE for embedded systems," in *Proc. 4th ACM Int. Conf. on Embedded Software (EMSOFT '04)*, Pisa, Italy, September 27–29, 2004, pp. 54–63.

Yuan, W. and Nahrstedt, K., "Energy-efficient soft real-time CPU scheduling for mobile multimedia systems," in *Proc. 19th ACM Symp. on Operating Systems Principles (SOSP '03)*, Bolton Landing, NY, October 19–22, 2003, pp. 149–163.

Zeng, H., Ellis, C., Lebeck, A., and Vahdat, A., "Currentcy: unifying policies for resource management," in *Proc. USENIX Annual Technical Conf.*, 2003, pp. 43–56.

Zeng, H., Ellis, C. S., and Lebeck, A. R., "Experiences in managing energy with ECOSystem," *IEEE Pervasive Comput.*, Vol. 4, No. 1, pp. 62–68, Jan. 2005. doi:10.1109/MPRV.2005.10

Zhu, Q. and Zhou, Y., "Power aware storage cache management," IEEE Trans. Comput., pp. 598–602, Vol. 54, No. 5, May, 2005.

Author Biography

Carla Schlatter Ellis is a Professor of Computer Science at Duke University. She received her Ph.D. degree in Computer Science from the University of Washington, Seattle in 1979. Before coming to Duke as an Associate Professor in 1986, she was a member of the Computer Science faculties at the University of Oregon, Eugene, from 1978 to 1980, and at the University of Rochester, Rochester NY, from 1980 to 1986. She is on the board of the Computing Research Association (CRA) and is Editor-in-Chief of ACM Transactions on Computing Systems. In the past, she served as a member of ACM Council and as Chair of the Special Interest Group on Operating Systems (SIGOPS). She is a member of ACM and USENIX. Her research interests are in operating systems, mobile computing, and sensor networks.